THE FIRST CAUSE

IPSUM ESSE SUBSISTENS AETERNUM

RETHINKING GOD IN A SCIENTIFIC AGE

by

A. J. ROWAN

THE FIRST CAUSE

IPSUM ESSE SUBSISTENS AETERNUM
Rethinking God In A Scientific Age
By A. J. Rowan

The question of God — the so-called "God hypothesis" — asks how we can possibly know that the First Cause exists. A philosophy student once asked, "When all is said and done, we can't really *know* that God exists, can we?" The answer is: we can, and we do.

Since the modern turn — from Hume's skepticism to Kant's revolution, and through the rise of the scientific worldview — it has become increasingly difficult to make intellectual room for God. We have grown doubtful not only of God's existence but even of our own place in the cosmos. Yet knowledge of God remains possible, both through *modal necessity* and through the evidence of sense experience.

Contrary to the claims of both Kant and Hume, our senses are not obstacles to knowing God; they are the very conditions that make such knowledge possible. They are the necessary and sufficient conditions through which we encounter the reality, necessity, and efficient causality of the First Cause.

Although it has been argued that *a posteriori* reasoning provides a basis for attaining knowledge of necessity and efficient causality through experience, such claims fail to clarify how this knowledge should be understood in light of Kant's *Critique of Pure Reason* and his development of the synthetic *a priori*. Nor do they adequately account for how knowledge of efficient causality and necessity—when derived from experience—can be employed in demonstrating the existence of God. The central question, therefore, is how such experiential knowledge might serve either to confirm or to challenge the God hypothesis, given the *a priori* status of causality and necessity in the post-Kantian framework.

If the *a posteriori* is understood in the Humean sense—as mere experience and custom independent of reason or the *a priori*—then for both Kant and, ostensibly, Aquinas, the *a posteriori* would be confined to the realm of sensibility and empirical observation. Yet Aquinas held no such Humean conception. Following Aristotle's critique of Plato's *noetic*, Aquinas rejected any division between experience and intelligibility. For him, sense and intellect were not opposed but ordered parts of a single act of knowing. This unity of cognition arose from the classical scholastic critique of the cognitive psychologies of Avicenna and Averroës, who, influenced by Neoplatonic thought, had separated the intellect from experience and thus fractured the continuity between the sensible and the intelligible.

Avicenna (*Ibn Sīnā*) and Averroës (*Ibn Rushd*), under the influence of late Neoplatonism, introduced a more radical separation between intellect and sense experience. They posited an *agent intellect* existing apart from the human knower—cosmologically emanated in Avicenna's account and conceived by Averroës as a single, shared intellect common to all minds. In their systems, human understanding depends upon mediation from this separate substance, rather than upon an intrinsic power of the soul itself. The consequence is a decisive bifurcation between experience and intelligibility, for the power of abstraction is no longer fully immanent within the human intellect but externalized beyond it.

Owing to the scholastic critique of this bifurcation between experience and intelligibility, Aquinas maintained that what is apprehended in singularity at the level of experience is abstracted in its universality at the level of intelligibility. In this synthesis, the intellect transcends both the Humean reduction of the *a posteriori* to mere custom and the Kantian confinement of knowledge to transcendental conditions. Thus, Aquinas entirely dissolves the modern dichotomy, restoring the unity of sense and intellect as a single act of knowing.

This book will therefore reexamine the arguments for God's existence through the lens of *efficient causality*, undertaking a critique and reformulation of Kant's synthetic *a priori* judgments concerning causation and modal necessity. In doing so, it seeks to dissolve the unwarranted divide between sensibility and intelligibility introduced by early modern philosophy.

From this renewed synthesis, a coherent argument will emerge— one demonstrating that the necessary existence of God can indeed be shown from a reformulated synthetic *a priori* knowledge of efficient causality and modal necessity.

Join me on this journey of discovery: a timeless pursuit of the unknown, and yet the knowable. If you have ever asked yourself, *Is it possible?*—or simply *Why or How?*—then this book is for you.

This volume marks Part I of a two-part series, in which we shall not only demonstrate the necessary existence of God, but also reveal that God is, in truth, the necessary First Cause of all existence. We are, together, in search of *the God proof.*

TABLE OF CONTENTS

LIST OF FIGURES

ACKNOWLEDGMENTS

The author wishes to acknowledge the thesis committee and those Professors of the Dominican School of Philosophy and Theology who provided the Thomistic background, resources, education and counsel and who have made possible the successful completion of this book. Specifically Professor John Hilary Martin, O.P., Professor Emeritus of Philosophy and Theology with particular emphasis in Medieval history, noetics, and cognition; Professor Stephen Ernest, S.V.D, Adjunct Professor of Philosophy with emphasis in epistemology and early modern philosophy; Professor Mark Damien Delp, Associate Professor of Philosophy with emphasis in Medieval logic and metaphysics; Professor Michael J. Dodds, O.P., Department Chair of Philosophy and Theology with emphasis in Divine action, philosophical anthropology, and philosophy of nature.

The author also wishes to acknowledge those Professors of the University of California Berkeley including Hannah Ginsborg for her skillful presentation of Kant's *Critique of Pure Reason* and the Kantian noetic; John MacFarlane Associate Professor of Philosophy who provided correspondence and needed support in researching various topics in Aristotelian modal logic and psychology; Brandon Fitelson's Assistant Professor of Philosophy and his emphasis in contemporary metaphysics; Soloman Feferman Professor of Mathematics and Philosophy at Stanford University and the late Donald Davidson and their presentation of contemporary logic and truth theory.

The author gratefully acknowledges his loving family, Melinda, Michael and John, for their patience and loving support. The author also wishes to express his deepest gratitude to his late sister, whose lifelong struggle with bipolar disorder has revealed in a profoundly human and personal way the tension between intelligibility and experience. Through her endurance of hallucinations and the shadows of schizophrenia, she has borne witness to the mystery of the mind's fragility and the soul's resilience—a living testament to the depths of the human spirit and the hope that meaning endures even in suffering. It is this experience that sparked interest in and the need for an integrated interdisciplinary approach between epistemology, the cognitive and brain sciences, logic, the philosophy of mind, and medieval philosophy.

Intelligibility disconnected from existence prevents assimilation to the things as they are in themselves. Such a disconnect does not accurately express, nor does it allow one to infer any conformity with existence whether in the case of a psychosis, nominalism, or even *the forms of the manifold of empirical intuitions* in the Kantian transcendental philosophy. For what is understood are imaginary conceptions in the case of a psychosis, names in the case of nominalism, or simply appearances or psychologically constructed phenomena in the case of Kant. Therefore, it is necessary to develop an interdisciplinary approach between the study of the mind and brain using the disciplines of epistemology, metaphysics, and the cognitive and brain sciences regardless of the proposed noetic. Although the proper object of metaphysics is being *qua* being, secondarily all sciences fall under the scope of metaphysics since metaphysics is the study of being. Epistemology and the philosophy of mind are informed by and inform the cognitive and brain sciences while the cognitive and brain sciences both limit and assess the validity of a given noetic theory.

ABBREVIATIONS USED ON REFERENCE LINES

A, a	Article
Ari	Aristotle
Bk	Book
Body, ad	Body
Ch	Chapter
CPR	Critique of Pure
Fn	Footnote
Lec	Lecture
Lsn	Lesson
Obj	Objection
OTC	On the contrary
Prt	Part
Q, q	Question
Rp,RpCn, Reply	Reply
Sct	Section
Thes	Thesis
V	Verse
Vol	Volume

Chapter 1

INTRODUCTION

Kant argued that metaphysics is a speculative science of reason, above experience, and that metaphysics rested on concepts alone.[1] However, Kant argued that *the crisis in metaphysics* was that metaphysics, resting upon concepts alone, had not entered upon the secure path of a science and as a result the procedure of metaphysics had "hitherto been a merely random groping, and, what is worst of all, a groping among mere concepts… how little cause have we to place trust in our reason, if, in one of the most important domains of which we would fain have knowledge, it does not merely fail us, but lures us on by deceitful promises, and in the end betrays us!"[2] Kant explains in the preface to the first edition of the *Critique of Pure Reason*, "for since the principles of which it [reason] is making use transcend the limits of experience, they are no longer subject to any empirical test. The battlefield of these endless controversies is called metaphysics."[3]

Arguing that metaphysics had failed in "resorting to principles which overstep all possible empirical employment, and which seem so unobjectionable but by this procedure human reason precipitates itself into darkness and contradiction,"[4] Kant introduced the Copernican

---

1 Immanuel Kant, *Critique of Pure Reason*, trans. by N. Kemp Smith (London: Macmillan, 1929), 21. All citations of *CPR* are from Immanuel Kant, *Critique of Pure Reason*, trans. by N. Kemp Smith (London: Macmillan, 1929) unless otherwise noted.
2 Ibid.
3 Ibid., 7.
4 Ibid.

revolution.[5] Following the examples of mathematics and natural science, Kant proposed considering their "essential features in the changed point of view by which they have so greatly benefited,"[6] and by way of *experiment*, Kant suggested to imitate their procedure.[7] Explaining the *Copernican Revolution*, Kant indicates: "Hitherto it has been assumed that all our knowledge must conform to objects. But all attempts to extend our knowledge of objects by establishing something in regard to them *a priori*, by means of concepts, have, on this assumption, ended in failure. We must therefore make trial whether we may not have more success in the tasks of metaphysics, if we suppose that objects must conform to our knowledge."[8]

---

5 Gilson in *Being and Some Philosophers* indicates that in the preface to Wolff's Ontology we read "Prime Philosophy (namely, metaphysics) was first laden by the Scholastics with enviable praise, but, ever after the success of Cartesian philosophy, it fell into disrepute becoming a laughing stock." Wolff, according to Gilson, held that since the time of Descartes there had been no metaphysics and what held claim to metaphysics had become unfashionable. Wolff was keenly aware of carrying on the work of the Scholastics but had misunderstood metaphysics reducing metaphysics to essentialism. Metaphysics under Wolff came to be understood in terms of essence rather than existence and essence and therefore under Wolff the essential elements of being as essence became ontology. Thus Kant's project against the dogmatists was against essentialism rather than scholastic metaphysics. By 1729, the Thomistic notion of being was completely forgotten. Wolff's notion of being and thus metaphysics was that of Francis Suarez, of the Society of Jesus, and Kant through his professor Franz Albert Schultz at the university of Koenigsberg had taken up the metaphysics of Wolff. Gilson explains that throughout the schools of Europe and particularly Germany, "metaphysics was Wolff and what Wolff had said was metaphysics." According to Gilson, Kant's "*Critique of Pure Reason* rests upon the assumption that the bankruptcy of the metaphysics of Wolff had been the very bankruptcy of metaphysics." Cf. Etienne Gilson, *Being and Some Philosophers* (Canada: PIMS, 1952), 114, 118-134.

6 Immanuel Kant, *Critique of Pure Reason*, translated by N. Kemp Smith (London: Macmillan, 1929), 22.

7 As an experiment, the Kantian shift in noetics made certain claims of the scientific Copernican revolution as being fundamentally a shift in epistemology without affirming that the scientific revolution was due to technological advance, mathematical modeling, and an inductive experimental method that limited speculative conjecture rather than a shift in noetics. The shift in noetics developed as a consequence of nominalism with the effect being a disconnect between empiricism and rationalism. By introducing transcendental idealism (i.e., an attempt at a middle ground between empiricism and rationalism grounded in nominalism), Kant further exasperated the problem of metaphysics by introducing a speculative *a priori* noetic as a method for doing metaphysics and logic detached from access to immediate existence through sense experience and assimilation. Of course as Aquinas writes, "the soul is not simply identical with the things it knows; for not stone itself, but its formal likeness exists in the soul. And this enables us to see how intellect in act is what it understands; the form of the object is the form of the mind in act" (*In De Anima* III, Lec 13 Sct 789). Therefore one can say "the soul in man takes the place of all the forms of being, so that through his soul a man is, in a way, all being or everything; his soul being able to assimilate all the forms of being--the intellect intelligible forms and the senses sensible forms" (*In De Anima* III, Lec 13 Sct 790). All citations of *De Anima* from *Aristotle's De Anima with the Commentary of St. Thomas Aquinas*, trans. by K. Foster and S. Humphries (New Haven: Yale University Press, 1951) unless otherwise noted.

8 Immanuel Kant, *Critique of Pure Reason*, translated by N. Kemp Smith (London: Macmillan, 1929), 22.

Kant concluded that sense objects must first conform to our knowledge. What Kant should have concluded is that the intellect is assimilated to external objects as the intellect apprehends those objects from experience or sensibility. However, for Kant the knowing subject became the determining factor of the object to be known: "this would work better with what is desired, namely, that it should be possible to have knowledge of objects *a priori*, determining something in regard to them prior to their being given. We should then be proceeding precisely on the lines of Copernicus' primary hypothesis."[9] Kant argued that Copernicus obtained success in explaining the movements of the heavenly bodies by making the spectator to revolve and the heavenly bodies to remain at rest.

Kant's Copernican Revolution held that experience itself is a species of knowledge that involves understanding; and understanding has rules which are intrinsic to the knowing subject prior to objects being given to the knowing subject, and therefore experience becomes *a priori* in "finding expression through *a priori* concepts to which all objects of experience necessarily conform, and with which they must agree."[10] Thus the objects of possible experience must necessarily conform to *a priori* concepts of the knowing subject giving metaphysics the secure path of science.

One of two main objectives of Kant's shift in metaphysics and noetics was to justify, in the face of Humean critical empiricism, the claims of science, placing science on a firmer foundation as a basis for having real knowledge of matters of fact.[11] Paul M. Churchland from the University of California, San Diego explained at the 2002 presidential address of the American Philosophical Association that Kant's agenda was to "vindicate an alleged class of synthetic *a priori* truths (e.g., geometry and arithmetic), and to explain in detail how such truths are possible."[12]

9 Ibid.
10 Ibid., 23.
11 W.T. Jones, *Kant and the Nineteenth Century* (New York: Harcourt Brace & Company, 1980), 65.
12 Elisabeth Radcliffe, e.d., *Proceedings and Addresses of The American Philosophical* Association (DE: The American Philosophical Association, 2002), 25.

Kant's second objective was to justify the moral insights of traditional religion against a mechanistic view drawn from Newtonian physics.[13] It should be noted that the second objective provided the main concern for each of the three critiques during Kant's critical period between 1781 and 1790. Before 1781, during Kant's pre-critical period, Kant was a Wolffian dogmatist but in reaction to Hume, Kant awoke from his "dogmatic slumber" or "Wolffian sleep" developing a framework described as transcendental idealism in an attempt to provide a philosophical justification for the sciences against Hume and to provide a basis for morality.[14] The underlining objective of the *Critique of Pure Reason* was to lay the groundwork of practical reason and morality, but the main purpose of the critique of pure speculative reason was to "revolutionize, in accordance with the example set by the geometers and physicists, the procedure that hitherto prevailed in metaphysics."[15]

As such, the critique of the faculty of reason, in respect of knowledge independent of experience, was intended to "decide the possibility or impossibility of metaphysics in general, and determine its sources, its extent, and its limits--all in accordance with principles."[16] Hume had stated in his Appendix to the *Treatise of Human Nature*, "nor is it in my power to renounce either of them, namely, that all our distinct perceptions are distinct existences, and that the mind never perceives any real connection among distinct existences."[17]

Arousing from a Wolffian and rationalist slumber, Kant realized the philosophical problem as explained by Gilson: "What are we to do with existence, if all our perceptions are distinct existences, and if the mind never perceives any real connection between them but

---

13 W. T. Jones, *Kant and the Nineteenth Century* (New York: Harcourt Brace & Company, 1980), 65.

14 Wolffian metaphysics had become a metaphysics without natural theology, it became a metaphysics of essence detached from existence. Thus Kant was correct to argue that metaphysics could not demonstrate the existence of God using the ontological argument since Wolffian metaphysics from which Kant was arguing was confronted with the task of finding a sufficient reason for the existence of the world and its connection to *esse* from mere concepts detached from experience because Wolffian metaphysics had detached essence from existence.

15 Immanuel Kant, *Critique of Pure Reason*, translated by N. Kemp Smith (London: Macmillan, 1929), 25

16 Ibid., 9.

17 Etienne Gilson, *Being and Some Philosophers* (PIMS, Toronto, Canada: 1952), 122.

only abstract essences. The solution was to suggest that the mind does not perceive such connections, it prescribes them."[18] Kant's method in putting metaphysics on a secure path of science was to show the scope and limits of reason.[19] For Kant, reason is our capacity for *a priori* knowledge. Kant undertook in *the Critique of Pure Reason* to investigate the scope and limits of *a priori* knowledge and this project had both a positive and negative aspect. The positive aspect was an attempt to show that we are capable of *a priori* knowledge and this aspect was opposed to Humean skepticism and skepticism in general.

Robert Paul Wolff commenting on Kant's perspective of Hume's critical approach explains that Hume's attack on causal inference undermined the use of intelligence as the production of pure concepts which derived nothing from experience and yet apply with universal validity to independent reality.[20] Although the natural solution for Kant would have been simply to limit the pure concepts to phenomena giving up the claim to noumena, Hume's attack however also undermined this position.[21] Hence Kant was forced to provide a general critique of the functions of human understanding and *a priori* knowledge forcing him to examine more closely the relation between understanding and sensibility.[22] According to Wolff, Kant's solution found in the *Deduction* and *Analogies* is the recognition that pure concepts are actually rules or the *schema* for the synthesis of a manifold of intuition.[23]

---

18 Ibid.

19 Thus, for Kant, in reducing metaphysics to the scope and limits of reason, metaphysics became noetics. Metaphysics is no longer being *qua* being but is reduced to analytic judgments and in dialectic there can be no metaphysical knowledge beyond the bounds of possible experience. Therefore, Kant's negative program that consisted in demonstrating the *limits* of *a-priori* knowledge concluded that God is beyond the bounds of possible experience; and in respect to dialectic, God was no longer considered cognitive. Contemporary philosophy for the most part agrees with Kant's dialectical perspective, and most Kantian work is at the level of the analytic and aesthetics. The Kantian Copernican shift was not only a shift in epistemological method, but it was a shift in categories where metaphysics became noetics, and as such, metaphysics was no longer able to critique noetics at the level of being *qua* being but rather noetics became the sole determining factor of what falls under the purview of metaphysics. The discovery of causes as such essential to metaphysics became the subject not of being but rather the subject of the *a priori* Kantian noetic. Rather than formal, efficient, final, and material causality falling under being as such, efficient causality became the proper subject of and became determined by the *a priori* for Kant.

20 Robert Paul Wolff, *Kant's Theory of Mental Activity* (Gloucester, Mass: Peter Smith, 1973), 31.

21 Ibid.

22 Ibid.

23 Ibid., 32.

Kant's *positive project* in effect was aimed at showing that we can have *a priori* knowledge, but the negative implication was that *a priori* knowledge is confined in the range of experience via the empirical intuition of appearances, and independent of direct experience of things-in-themselves. In other words, the *positive project* of the possibility of *a priori* knowledge implied the *negative project* of setting the limits of reason where *a priori* knowledge extends only as far as a range of experience via the empirical intuition of appearances. However, this *negative limitation* of speculative reason likewise implied a *positive advantage*. In the negative case of limiting speculative reason, Kant in arguing for an "unavoidable ignorance of things in themselves, limiting all that can be theoretically known to mere appearances," Kant believed he was also providing a *positive advantage* for the doctrine of moral judgment, God, and the simple nature of the human soul limiting such knowledge to the realm of faith.

The implication of Kant's *positive advantage* in limiting the scope of speculative reason was one of relegating knowledge of God and of God's existence to faith, and as such, knowledge of God's existence cannot be considered cognitive in a Kantian sense. This does not mean one cannot think about God, but knowledge of God's existence cannot conform to the *form of empirical intuition* and *rational judgment*. The *form of empirical intuition* can refer to either the "formal features or structure of the objects intuited equivalent to 'form of appearances' or the manner or mode of intuiting."[24] The *form of appearances*, where an *appearance* is an undetermined object of an *empirical intuition*, is that which so determines the *manifold of appearance* that it allows of being ordered or related to one another in experience.

The form of appearances or the form of empirical intuition is where a form is a feature of the appearance or empirical intuition in virtue of which their elements are viewed as ordered or related to one another in possible experience. God's existence does not conform to objects that are actually given in possible experience in contrast to an

---

24 Henry E. Allison, *Kant's Transcendental Idealism* (New Haven: Yale University Press, 1983), 106.

object that is merely conceived through the form of empirical intuition. Therefore, Kant argued that our assumptions of God, freedom, and immortality are permissible only by faith because speculative reason only extends to objects of *possible experience* according to a given form that orders possible experience in conformity to some epistemic condition or rule.[25] One can only make coherent knowledge-claims about objects that are spatial and temporal, because space and time are the pure forms of human intuition. Although we find that mathematics and physics yield genuine knowledge because they are concerned with objects of possible experience, this is not the case with God's existence or knowledge-claims about God. Although God's existence is conceivable, God's existence does not conform to epistemic conditions and is not an object of *possible experience* in Kant's noetic, and therefore, knowledge of God's existence and knowledge-claims about God became noncognitive being relegated to either opinion or faith existing independent of pure reason and *actual experience* after the Kantian turn.

To propose that we can have certain knowledge of God's existence in virtue of demonstration or speculative reason would deprive speculative reason of any pretension to transcendent insight because to arrive at transcendent insight speculative reason must make use of principles that only extend to objects of *possible* experience and *a priori* knowledge. However, *a priori* knowledge extends only as far as a *range of experience* via *empirical intuition of appearance* in virtue of which a knowledge of God would not be considered cognitive, but this denial

---

25 According to Aquinas Aristotle's distinction between speculative and practical reason was a distinction between intellectual pursuits directed toward the cause of truth in and for itself and pursuits requiring practical knowledge or skill such as building a house where truth is pursued in relation to action (cf. *In Meta*. II.2.290). Of course, speculative reason for Aristotle did not extend to objects of *possible experience* according to a given form that orders possible experience in conformity to some epistemic condition or rule. As will be seen, for Aristotle first principles were apprehended from experience inductively such that first principles had their foundation in being itself or sensible things. Aristotle writes: "Now induction is of first principles and of the universal and deduction proceeds from universals. There are therefore principles from which deduction proceeds, which are not reached by deduction; it is therefore by induction that they are acquired" (*Nich. Eth*. VI, 1139b19-1139b35). All citations of *Nich. Eth*. from *The Complete Works of Aristotle The Revised Oxford Translation* Vol II ed. by Jonathan Barnes *Nicomachean Ethics* trans. by W. D. Ross and revised by J. O. Urmson (New Jersey: Princeton University Press, 1984)).

of a knowledge of God and particularly of God's existence also has the positive advantage according to Kant of making room for faith: "I have therefore found it necessary to deny knowledge [of God], in order to make room for faith. The dogmatism of metaphysics, that is, the preconception that it is possible to make headway in metaphysics without a previous criticism of pure reason, is the source of all that unbelief, ... which wars against morality."[26]

However, the negative aspect of Kant's positive advantage denies the possibility of a speculative knowledge and demonstration of God's divine action and divine causality thus limiting speculative knowledge of God's existence to a theory of faith independent of reason and possible experience since *a priori* knowledge is confined to the range of possible experience via the empirical intuition of appearances. Divine action, divine causality, and God's existence are simply outside the range of the empirical intuition of appearances, and thus outside possible experience, and therefore beyond the limits of pure reason. It is this fundamental shift in epistemology and the Kantian understanding of causality, universality, and necessity that this thesis attempts to address.

Although Aquinas agrees with the limitations of human reason to obtain a natural knowledge of God and often repeats that "God can be thought not to be not because of God's imperfections or uncertainty but because of the weakness of our intellect," yet Aquinas maintains that because of this limitation, "one cannot behold God Himself *except through His effects that lead to a knowledge of His existence through reasoning*" (emphasis added).[27] Although it is clear that there is a limitation, Aquinas

---

26 Immanuel Kant, *Critique of Pure Reason*, trans. by N. Kemp Smith (London: Macmillan, 1929), 29.

27 *SCG* I 11.4; Likewise, Aquinas explains: "Creatures fail to represent their creator adequately. Consequently, through them we cannot arrive at a perfect knowledge of God. Another reason for our imperfect knowledge is the weakness of our intellect, which cannot assimilate all the evidence of God that is to be found in creatures. It is for this reason that we are forbidden to scrutinize God's attributes overzealously in the sense of aiming at the completion of such an inquiry, an aim which is implied in the very notion of overzealous scrutiny. If we were to act thus, we would not believe anything about God unless our intellect could grasp it. We are not, however, kept from humbly investigating God's attributes, remembering that we are too weak to arrive at a perfect comprehension of Him" (*De Veritate* II, a. 5, a. 2, reply 11). All citations of *De Veritate* are from *De Veritate* vol. I trans. by Robert William Mulligan, S. J. (Chicago: Henry Regnery Co., 1952); vol. II trans. by James V. McGlynn, S.J. (Chicago: Henry Regnery Co., 1953); vol. III trans. by Robert W. Schmidt, S.J. (Chicago: Henry Regnery Co., 1954) unless otherwise noted.

explains that it is exactly because of this limitation that one cannot behold God by natural reason except through His effects. However, in denying the knowability of divine action and divine causality from actual experience, Kant's critique of pure reason disqualified one's ability to know from experience God's existence or any knowledge about God from his effects. Hence one's natural knowledge of God is entirely limited to opinion and a nominal understanding of God and of God's existence rather than faith where faith is understood to be an unwavering assent that something is true having been persuaded from reason and/or one's experience and thus moved by the will to assent.[28]

In limiting speculative reason to analytic and synthetic *a priori* judgments and in limiting the scope of *pure* reason, Kant successfully

---

28 Faith for Aquinas is an unwavering assent that something is true in judgment by the influence of the will (cf. (*De Veritate* II, q. 14, a. 1, body)). The understanding or judgment is moved by the will and thus judgment is moved or determined by those things proper to the will and external to itself, by some evidence such as something experienced or the testimony or eye-witness accounts of others who have experienced, or something fitting, or something useful and thus the understanding is held captive by the will that something is true. In contrast, doubt is the inability to either affirm or deny that something is true or false and although opinion may lean toward one side or the other it always doubts and thus opinion is not assent that something is true. In the case of faith, the understanding still thinks discursively and inquires about the things which it believes, even though its assent to what is true is unwavering. Opinion likewise still thinks discursively but opinion always doubts unable to resolve the conflict between two sides of a contradictory proposition. Aquinas explains the distinction between doubt and opinion: "For, sometimes, it [the intellect or judgment] does not tend toward one rather than the other, either because of a lack of evidence, as happens in those problems about which we have no reasons for either side, or because of an apparent equality of the motives for both sides. This is the state of one in doubt, who wavers between the two members of a contradictory proposition. Sometimes, however, the understanding tends more to one side than the other; still, that which causes the inclination does not move the understanding enough to determine it fully to one of the members. Under this influence, it accepts one member, but always has doubts about the other. This is the state of one holding an opinion, who accepts one member of the contradictory proposition with some fear that the other is true" (*De Veritate* II, q. 14, a. 1, body).

Those who have opinion may lean toward one position or another but they remain uncertain, but belief is assent with certainty that something is true even while in parallel the intellect discursively inquires about the things which it believes. Aquinas explains the distinction between assent and opinion: "For we are not said to assent to anything unless we hold it as true. Likewise, one who doubts does not have assent, because he does not hold to one side rather than the other. Thus, also, one who has an opinion does not give assent, because his acceptance of the one side is not firm. The Latin word *sententia* (judgment), as Isaac and Avicenna say, is a clear or very certain comprehension of one member of a contradictory proposition. And *assentire* (assent) is derived from *sententia*. Now, one who understands gives assent, because he holds with great certainty to one member of a contradictory proposition" (*De Veritate* II, q. 14, a. 1, body). Although in moral judgment, Kant left the notion of what is fitting or useful, Kant in removing from possible experience any evidence of God from his effects, left no room for assent but only opinion in such matters. In removing a natural knowledge of God from possible experience thereby making room for faith, Kant's theory of faith relegated belief in God's existence and any natural knowledge about God to opinion.

limited knowledge of God's existence to practical reason or what is morally fitting and to a Kantian theory of faith considered independent of experience as well as independent of universality and necessity. In defining universality and necessity in terms of the *a priori*, logical certainty was limited to the range of possible experience via empirical intuition of appearances; knowledge of God was thus limited to and rested upon the subjective grounds of the certainty of one's own personal and individual moral sentiment independent of experience, necessity, or a natural knowledge of God through His effects, thus disqualifying a natural knowledge of God's existence through a critique of pure reason.[29]

Although a number of reasons explain why God exists, or why one has freedom, or reasons for an immortal soul, such reasons cannot be addressed by synthetic *a priori* judgments and therefore can never be of a universal or necessary nature nor can such reasons have the quality of logical certainty as understood by Kant. However, God, the immortality of the soul, etc. *are needed as things to act morally* and are to be *assumed by faith*. On the positive side, by limiting knowledge of God or any metaphysical demonstration of God's existence insofar as such knowledge might be based upon speculative reason, Kant believed he had made room for faith while expressing the full scope of practical reason. The *Copernican shift* introduced by Kant was an epistemological turn limiting the bounds of speculative reason thus introducing a limitation on what can be known by speculative reason having the consequence that God is noncognitive and God's existence cannot be argued for from speculative reason.

In response to the mind never perceiving any real connection between what is perceived and what exists following Hume's critique, Kant held that rather than the mind perceiving such a connection, the mind conditions or prescribes the connection. The fundamental issues that drove the question of a connection between perception and existence were the issues of substance and causality. Kant wished to

29 *CPR*, A 829/B 857.

justify universal and necessary causal connections and since such connections could not be perceived, they must therefore be universally and necessarily prescribed. An *a priori* basis for substance and causality allowed one to do science and metaphysics, but to provide this *a priori* ground, Kant argued that one can only obtain from possible experience or existence the notion of substance or causality in virtue of the *manifold of appearance* as determined by *the form of empirical intuition* in *synthetic judgments*. However, this also limited the scope of speculative reason to the range of the *empirical intuition of appearances* such that God's existence and knowledge about God fell outside of what can be known from speculative reason, and the connection between the knowability of efficient causality and of God's existence from efficient causality and necessity was dissociated in philosophical discourse.

Kant, following Hume, maintained that the *a posteriori* synthetic relation is only empirical and the representation is therefore never *a priori* and thus the *a posteriori* can never lead to necessity or universality. Although Thomists have proposed that *a posteriori* judgments provide a basis for knowledge of efficient causality, this fails to explain how knowledge of efficient causality can be understood in terms of necessity and universality, and still be apprehended from experience given Kant's *a priori* and *a posteriori* distinction. Nor does such a position explain the relation between the *a priori* in Kant and how knowledge of efficient causality and necessity from experience can even be used in the second and third demonstrations within the Five Ways used by St. Thomas Aquinas given the *a priori* nature of causality and necessity after Kant.

The *purpose of the thesis* will therefore reconsider the argument for God's existence based on a knowledge of efficient causality in Aquinas's second way and necessity in Aquinas's third way by a critique and reformulation of Kant's *a priori* knowledge of synthetic judgments about efficient causality and necessity. Kant agrees with Hume's insistence that there is nothing in existence itself linking an effect to its cause but disagrees with Hume's assertion that any such association is not by reason but arrived at only from custom. Kant maintained against Hume that there still remains a *universal and necessary* connection between a cause and its effect whereby pure reason judges *a priori* the causal maxim that states "whatever begins to exist proceeds from some cause."

Kant thus shifted the issue of *necessary universal causality* from existence or experience to one of synthetic *a priori* judgments where "*a priori*" refers to ways of knowing a judgment such that the judgment is necessarily and universally true. The problem for Kant was to show that one could know synthetic judgments of efficient causality *a priori* while continuing to accept Hume's position that there is nothing in existence itself that can be experienced linking an effect to its cause. Hume maintained that contiguous resemblance and custom allowed one to believe and thus infer a necessary causal connection.

Against the synthetic *a priori* transcendental philosophy of Kant and in opposition to Hume's empirical *a posteriori* development and critique of reason, *the work to be done* includes showing that the certainty of God's existence can be known by an *a priori* connection between an effect and its cause as apprehended from experience independent of the Kantian transcendental in Aquinas's second and third demonstrations in the Five Ways. This work will be accomplished by a) providing a historical critique showing that the Kantian transcendental was a historical and cultural development founded upon a false dichotomy between sensibility and intelligibility, b) providing an alternative to the Kantian transcendental philosophy of the synthetic *a priori*, and c) demonstrating God's existence from a synthetic *a priori* knowledge of motion, efficient causality, and necessity.

The chapter immediately following this introduction will discuss the historical developments of the transcendental deduction. We will find that similar issues between sensibility and intelligibility or between body and soul or body and mind was an ongoing topic for philosophy at least since the pre-Socratics and Plato with Aristotle introducing the hylomorphic principle as a means to bridge the disparity between intelligibility and sensibility.

Similarly, we find that the Aristotelian solution to the mind-body problem was simply no longer accessible to the modern period after the introduction of nominalism and Cartesian dualism, after the rejection of any reliable knowledge of experience from proper (secondary qualities) and secondary (primary qualities) sensibles following Berkeley, and the rejection of essentialism and natural necessity as well as any knowledge of substance and efficient causality from sensibility following Hume's critique.

Kant thus found himself, after Hume and the developments in cognitive psychology and perception during the early modern period, in need of a solution to explain the relationship between sensibility and intelligibility. This chapter focuses upon the issues involved and the development of such a solution by Kant often termed the Copernican revolution. In opposition to all other accounts and the various proposed solutions prior to Kant, the uniqueness of Kant's solution was to develop a metaphysical and epistemological account that would reorder the entire way one perceives history, space, time, and reality itself.

The *third chapter* will emphasize the problem of the synthetic *a priori* and discuss the solution proposed by Kant. The *fourth chapter* explains the mind-body relationship understood by Aquinas drawing upon both the hylomorphic principle of Aristotle and the further development of the Dionysian principle of hierarchy used to address developments in Galenic physiology and that of Avicenna and what is today called cognitive psychology. This chapter will discuss and establish a reformulation of the synthetic *a priori* independent of the transcendental philosophy. The *fifth chapter* will argue for knowledge of efficient causality taken to be universal and necessary and therefore synthetic *a priori* while being independent of the Kantian transcendental philosophy based on the reformulation of the synthetic *a priori*.

Knowledge of efficient causality and necessity were key issues that concerned the early modern period. The way in which the early modern period addressed these fundamental issues led to the development of the Kantian turn in metaphysics and epistemology. The Kantian development of a new metaphysics of theism denied a natural knowledge of God's existence as apprehended from God's effects as expressed in and by nature or experience or sensibility, and thus denying a natural knowledge of God from possible experience, matters of faith were detached entirely from reason.[30] For early modern philosophy, the questions became, how can a causal *connexion* be abstracted from contingent experience given the unreliability of sensation where one only finds regularity, and how can this contingent perception possibly be both universal and necessary and thus *a priori*?

Chapter five will thus develop contrary to the *metaphysical deduction* that the categories are not transcendental rules, which underlie and make possible actual empirical synthesis, but *instead universal concepts are combined and separated in judgment following the law of contradiction allowing for synthetic a priori judgments insofar as such judgments have unity, universality, and necessity and are abstracted from sensibility.* Rather than the categories being antecedent to all experience as Kant would have it, *the categories are themselves ontologically grounded in natural substances and natural necessity following an essentialist awareness of causality, and it is these categories that underlie and make possible the*

---

30 For historical developments concerning this see Gordon E. Michalson, Jr., *Lessing's 'Ugly Ditch': A Study of Theology and History* (United States: The Pennsylvania State University, 1985). Stewart and Wright indicate "a historical case can be made that Hume actually had an influence on [the] Kierkegaardian conception [of faith], at least indirectly. For it can be shown that his discussions of belief in the Treatise and Enquiry deeply influenced Hamann and Friedrich Heinrich Jacobi in their conceptions of faith; and Kierkegaard knew both Hamann and Jacobi. For more on Hume and Hamann, see Philip Merlan, 'From Hume to Hamann,' Personalist 32 (1951): 11-18; 'Hamann et les dialogues de Hume,' *Revue de metaphusique* 59 (1954): 285-9; 'Kant, Hamann-Jacobi and Schelling on Hume,' *Rivista critica di storia di filosofia* 22 (1967): 3, 43-51." Cf. Manfred Kuehn, "Kant's Critique of Hume's theory of faith" in *Hume and Hume's Connexions*, ed. Stewart and Wright, (Pennsylvania: Pennsylvania State University, 1995), 239, 253 for a discussion of the "Modernist Protestant and Catholic" agreement with Hume's theory of faith. For a certain caricature of the contemporary Barthian view of faith one can see A. N., Prior, "Can religion be discussed?" in *New essays in Philosophical theology*, ed. A. G. N. Flew and A. C. MacIntyre (London: 1955): 1-11.

actual empirical synthesis at the level of perception and judgment. Freddoso describes this essentialist position: "it is of the essence or, as the term 'natural necessity' suggests, of the nature of these causes to act in this way and bring about this effect when unimpeded. Their producing such-and-such an effect in such-and-such circumstances flows from their being the sorts of things and, hence, the sorts of causes they are. They are what they are; they can be no other. And their action follows from what they are."[31]

It will also be argued against the *transcendental deduction*, that synthesis is not made possible by the Kantian transcendental categories but by cognitive psychology and intellectual judgment. Simple apprehension is where the proper formal object of the intellect is the abstract essence of sensible things, which is a quiddity or nature existing in corporeal matter. The proper formal object of the intellect depends upon the object of the senses, since the intellect understands by abstracting the intelligible species from matter. What is abstracted from matter is the universal where the intellect understands the universal directly through the intelligible species, and the intellect understands indirectly the singular by turning to sensibility by means of perception and the faculties associated with cognition.

The first operation of the intellect or simple apprehension apprehends what a thing is. It is that act of the intellect by which the intellect forms simple concepts of things by understanding the "whatness" or *quiddity* or the *essence* of each one of them. William Wallace explains: "simple apprehension is the first act of the intellect whereby it simply grasps what a thing is, i.e., its essence (*essentia*) or quiddity, without affirming or denying anything of it. In apprehending a quiddity the intellect forms within itself the formal concept... The concept is both intellectual knowledge, 'that which' (*id quod*) is understood, and the means 'by which' (*id quo*) the thing known is understood. These two features of the concept are spoken of as

---

31 Alfred J. Freddoso, "The Necessity of Nature," *Midwest Studies in Philosophy*, XI (1986) : 223.

the objective concept and the formal concept respectively."[32] Wallace continues by explaining that the objective concept consists of both extension and comprehension where comprehension is intention.[33] The concept is thus the internal representation of a thing's essence or quiddity, and the oral or written term or word is the external sign of the concept.[34] Wallace explains, "sense knowledge possesses the perception of the single, concrete, material object from which the nature or quiddity of the object is abstracted leaving aside all individuating notes that characterize it in its singularity."[35]

This process results in the intelligible species "which gives rise to the abstract, universal idea or concept of the object."[36] The intelligible species is not the concept itself, but it is that from which the concept is formed. The abstract, universal idea or concept formed or conceived (*conceptus*) is a formal sign of the nature or essence grasped in the object.[37] Wallace explains, "like the senses, the intellect does not know the concept as such, but rather knows and understands the object as it is in reality through the concept."[38]

The intellect thus understands the universal whereby the many are one (*id quod natum est praedicari de pluribus; unum (natura una) in multis et de multis*) *independent* of the Kantian *transcendental*. Synthesis of a given subject and object on the other hand occurs at the level of the second operation of the intellect by which the intellect combines and

---

32 William A. Wallace, O.P., *The Elements of Philosophy* (New York: ALBA House, 1977), 14.

33 Ibid.

34 Ibid.

35 Ibid., 72.

36 Ibid.

37 In modern scholastic terminology, the phantasm or perception of a concrete material extra-mental object is obtained through sense knowledge. The agent intellect (the active power of the intellect) extracts from the phantasm the nature or quiddity or essence (*essentia*) of the object, leaving aside all individuating conditions resulting in the intelligible species. The agent intellect then impresses the intelligible species or impressed species upon the possible intellect giving rise to the abstract, universal idea or concept of a given object. The intellect knows and understands the extra-mental existence of the external object as it is in existence through this universal idea or concept. The intelligible species arising in the agent intellect is the impressed species or the internal word and the concept formed in the possible intellect is the expressed species or external word. Cf. William A. Wallace, O.P., *The Elements of Philosophy* (New York: ALBA House, 1977), 14, 72.

38 William A. Wallace, O.P., *The Elements of Philosophy* (New York: ALBA House, 1977), 72.

separates, independent of the Kantian transcendental (*In Meta.* Book VI, L.4, 1232). Necessity can be attained from experience, and when abstracted as a universal, the intelligible species allows for the possibility of a synthetic *a priori* judgment where the predicate is not contained in its subject and the judgment is both universal and necessary. The universal, taken as the abstract essence of sensible things predicated of many, furnishes the required universal identity for conformity in the second operation of the intellect where conformity, combination, and separation occur whereby knowledge of *a priori* synthetic efficient causality and necessity from experience is possible.

The *sixth chapter* will argue for God's necessary existence from knowledge of synthetic *a priori* judgments of efficient causality following Aquinas's arguments in the second and third ways. Universal concrete concepts of natural or absolute necessity and efficient causality as well as other associated intrinsic irreducible causal powers and tendencies that are tied to substance and ontologically grounded in natural substances allow for a causal connection or active power between an effect and a given cause by which one can argue for God's necessary existence from efficient causality and necessity. Thus from the arguments from efficient causality and necessity, a synthetic *a priori* knowledge of God's existence will be demonstrated.

Through a reformulation of Kant's *a priori* knowledge of synthetic judgments about efficient causality, God's existence will be demonstrated from efficient causality and necessity. *Against Kant, it will be argued* in chapter four through six that the certainty of God's existence is itself grounded in an *a priori* connection between an effect and its cause both in judgment and experience in Aquinas's second and third demonstrations in the Five Ways independent of the Kantian transcendental. Independent of the transcendental exactly because one can attain universality and necessity abstracted from experience, and thus the *a priori* synthetic proposition can be abstracted from experience while being independent of experience once abstracted. Such synthetic proposition are thus independent of the transcendental philosophy in the sense of being independent of the rules for the imaginative synthesis of the manifold or simply independent of the rules or schemata some of which are empirical and others are *a priori*. The *a priori* nature of such concepts are not dependent upon empirical or *a priori* schemata but upon the process of abstraction from sensibility of the categories of being reflected by existence as existence is and not as it is imagined to be or as logically determined by the schemata.

The *final chapter* will conclude that universal and necessary *a priori* synthetic propositions are possible and Kant's program was in fact an unnecessary development of early modern philosophy. Kant argued in the Critique: "the proper problem of pure reason is contained in the question: How are a priori synthetic judgments possible? That metaphysics has remained in uncertainty and contradiction is due to this problem and because no one has considered the distinction between analytic and synthetic judgments. Upon the solution to this problem, or upon a sufficient proof that the possibility which it desires to have explained does in fact not exist at all, depends the success or failure of metaphysics."[39] The thesis will conclude using the historical critical method and a proper reformulation of the synthetic *a priori* that the *possibility* of a solution which the Kantian transcendental philosophy desired to explain does in fact exist, but simply not in the manner in which the Kantian solution proposed.

Kant left one with two options, either the transcendental solution or no solution at all. However, this is a false dichotomy since there are other possible solutions to the problem of the *a posteriori*/synthetic *a priori* distinction beyond those which Kant and Hume considered. The thesis will conclude that a reformulated synthetic *a priori* solution is a far more appropriate response to the Kantian synthetic *a priori* than proposing an *a posteriori* solution when it comes to resolving issues of the Kantian transcendental philosophy.

---

39 Immanuel Kant, *Critique of Pure Reason*, trans. by N. Kemp Smith, (London: Macmillan, 1929), 55.

The Kantian program was introduced by the debates between the British Empiricists and the Continental Rationalists, which were historically and fundamentally flawed, introducing a false dichotomy between sensibility and intelligibility. The debate introduced a mistaken disparity between the ontological status of causality and an empirical knowledge of causal necessity as well as between sensibility and intelligibility. Kant as a Continental Rationalist, awakened from his Dogmatic slumber having obtained empiricist Newtonian leanings, and Hume as a British Empiricist, necessarily followed the disparity between sensibility and intelligibility developed by early modern philosophy in their respective solutions. Although Kant's solution was an attempt at a compromise between British Empiricism and Continental Rationalism after Hume, transcendental idealism was essentially a Continental Rationalist development in reaction to the British Empiricism of Hume.

Having developed a cogent argument that universal and necessary knowledge of efficient causality can be attained from experience independent of the transcendental philosophy, it will be concluded that one can demonstrate God's existence from knowledge of modal necessity and efficient causality laying a philosophical foundation for an integrated model for an intelligible and sensible knowledge of God's existence through a reformulated synthetic a priori solution.

Chapter 2

THE HISTORICAL DEVELOPMENT OF TRANSCENDENTAL IDEALISM

Kant was trained in philosophy at the University of Konigsberg under Martin Knutzen, Professor of Mathematics and Philosophy. Knutzen taught the metaphysics of Leibniz, which had been made popular by Christian Wolff. The professor was also interested in the mathematical physics of Newton.[40] It is here that Kant was confronted with the contradictions and various conflicts between the continental rationalism of the *Dogmatists* including Leibniz and Christian Wolff and the British empiricism of Newton and Newton's defender Samuel Clarke referred to as the *Skeptics*. Kant was influenced heavily by the Clarke-Leibniz correspondence of 1715 to 1716 that provided Kant with the various arguments of both the Dogmatists and the British empiricists.[41] The fundamental issue at hand in the debate was the notion of space and time. The commentator Robert Paul Wolff explains: "the analysis of space as a set of relations gives to it an ontological status which is consistent with the rationalist metaphysics of the continent, but a doctrine of absolute space seems to be required for the new physics. This dilemma was the starting point for Kant's early studies in philosophy and science."[42]

---

40 Robert Paul Wolff, *Kant's Theory of Mental Activity*, (Gloucester, Mass: Peter Smith, 1973), 9.
41 Ibid., 4, 8.
42 Ibid., 8.

In Kant's earliest publication *Thoughts on the True Estimation of Living Forces* (1747), Kant attempted to defend the Dogmatist metaphysics of space and time as being relational as held by Leibniz but was unsuccessful. Here Kant attempted to justify the nature of space in terms *of causal interactions of unextended* substances arguing that substances must interact in order for them to be spatially related, and without some force acting outside of themselves, substances would have no extension and therefore no space.[43] Against Clarke and Newton, Leibniz had maintained that space is relational and Leibniz denied the existence of absolute space-time and the possibility of a void.[44] For Leibniz positions and dates are to be defined in terms of the objects given at that date or position, and where no substances exist, there can be no empty space nor empty time.

Having roots in Bacon's inductive experimental method and his critique of hypothetical deductive speculation, Newton and the British empiricists in general, eschewed all "hypothesis" and were suspicious of the Wolffian and Leibnizian metaphysics.[45] Newton in contrast to Leibniz viewed space and time as independent of objects and infinite based upon his own conception of physics. Therefore Newton concluded that absolute positions in space and time are not dependent upon objects which occupy space and time, and second that time, void of a referent, and empty space are possible and Newton felt that this theory of space and time was required by his view of motion.[46] During this early period, Kant attempted to find a satisfactory version of

---

43 Ibid., 9. It should be noted that for the *via antiqua* "substance" was not accidental (e.g., quantity or extension), but what existed was the composite of matter and form that consisted of a composite substance-accident being, and accidents inhered in or were related to the composite substance-accident existence. For example in the case of the composite substance of man in motion from point a to point b, the subject man would be accidentally related to either point a or b in virtue of the accident of place or location, and accidentally related in time between point a and b. Likewise, the accident of color inhering in a natural substance, for example the redness of a red rose, where if one grants that the physiology is correct, sight absorbs the photons of particular wavelengths caused by light reflecting upon the redness inhering in the rose so that the wavelengths change the shape of pigments in the case of rods or cones within the retina consistent with the color that acts upon visual perception as reflected from a given object.
44 Robert Paul Wolff, *Kant's Theory of Mental Activity*, (Gloucester, Mass: Peter Smith, 1973), 4.
45 Ibid.
46 Ibid.

Leibnizian metaphysics, but failed to do so resulting in a shift toward empiricism.[47]

At the end of Kant's early period, Kant published an essay adopting the British empiricist position of Newton entitled *On the First Ground of the Distinction of Regions in Space* where Kant draws upon an argument from Clarke in the Leibniz-Clarke correspondence to argue against the relational theory of space, which is believed to have devastated the Leibnizean metaphysics of space and time. With this essay Kant rejected the Dogmatic position of space and time and embraced the empiricism of Clarke and Newton and became a defender of British empiricism and particularly the Newtonian mathematical physics. Robert Wolff explains, "with this essay, Kant came to the end of his attempts to find a defensible version of rationalist metaphysics. In his next published work, the *Inaugural Dissertation* of 1770, he struck out in a radically new direction, and began the fundamental rethinking of philosophical issues which resulted in the *Critique of Pure Reason*."[48]

Although it is true that Kant struck out in a radically new direction under the influence of British Empiricism, this new direction remained essentially a Continental Rationalist development embracing Hume's critique and the early modern developments in psychology and perception as well as being influenced by early modern nominalism. In any case, the *Dessertation* was a pivotal shift for Kant in that he realized that the primary issue between rationalism and British empiricism was an issue between sensibility and intelligence. The full title of the *Dissertation* is *On the Form and Principles of the Sensible and Intelligible World*. Kant states the purpose as being an exposition of the concept of the world as a complex of substances, and the problem to be solved was the paradox that it seems *impossible* and yet *necessary* to form an adequate representation of a complex of substances.[49]

---

47 Ibid., 9.
48 Ibid., 10-11.
49 Ibid., 12.

Drawing upon distinctions in Wolffian logic and metaphysics, Kant's solution was to draw the classical distinction between two modes of representation, sensibility (ontology) and intelligence (epistemology). In medieval terminology, a similar distinction was made between real being and intentional being, a distinction between *modus essendi* and *modus intelligendi*. Augustine likewise made a similar distinction, as did earlier philosophers such as Plato.[50] However, the approach used to bridge the distinction between the sensible and intelligible as developed by Aristotle against Plato and recovered in the medieval *via antiqua* and used with qualification in the medieval *via moderna* was entirely different than and *opposed to* the 18th century transcendental idealism proposed by Kant.[51]

The significant difference between each of these epistemic conceptions is that objects external to a given knower exist independent of the knower and can be known by recourse to the external objects.[52]

---

50 Augustine says: "I regret that I said that there are two worlds, one the object of sense, the other the object of intellect--not because this is not true, but because I said it as though it were an original idea, when in fact it had been previously pointed out by philosophers, and because this manner of speaking is not usual in Holy Scripture" (St. Augustine, *Retractationum*, I, 3 (PL 32:588)).

51 The term *via antiqua* applies to Thomists, Scotists, and others. As far as the *via moderna* as Boehner explains, "The 'Nominales' (in the mediaeval sense) constituted the *via moderna*, which was not so much a school as a trend of thought" (cf. Philotheus Boehner, *William of Ockham: Philosophical Writings*, (Indianapolis: Hackett Pub. Co., 1990), li). Courtenay indicates that in the fourteenth-century even Ockham's closest followers, such as Adam Wodeham, were critical of Ockham's epistemology: "Ockham's removal of sensible and intelligible species in his explanation of the acquisition of knowledge was rejected by most of his English contemporaries, as was his definition of the object of knowledge (William J. Courtenay, "The Academic and Intellectual World of Ockham" in *The Cambridge Companion to Ockham*, ed. Paul Vicent Spade, (Cambridge: Cambridge University Press, 1999), 28). A similar sentiment continued in the 14th century at Oxford where "Thomas Bradwardine attacked Ockham's views on grace and justification as being Pelagian," and due to a *countercurrent* of realism against Ockham's nominalism (Courtenay, 28). However, in fifteenth century Paris the "reemergence of thought and division of faculties of arts based on the Aristotelian commentaries of Albert, Thomas, and Giles on the one side (the *via antiqua*) and a preparation based on the commentaries of Ockham (the *via moderna*), resulted in Ockham becoming "textually wedded to the 'modern' approach and an important authority of the *Nominalistae* at Paris and universities in Germany. By the end of the fifteenth century Ockham's name had become identified with a school of thought, and 'Ockhamist' took its place alongside 'Thomist,' 'Albertist,' and 'Scotist'" (Courtenay, 28-29).

52 It seems that Kant would have held that the medieval "realist" *via antiqua* and the medieval "nominalist" *via moderna* were both *transcendental realist* accounts just as the rationalism of Leibniz, the idealism of Berkeley, and the empiricism of Hume and Newton were all reducible to what Allison refers to as *transcendental realism* following Kant (cf. Robert Paul Wolff, *Kant's Theory of Mental Activity*, (Gloucester, Mass: Peter Smith, 1973), 16-25). Transcendental realism holds that "these modifications of our sensibility are considered to be self-subsistent things, that is, mere representations are treated as

Kant's epistemic conception reversed this order such that all that can be known are appearances that have some reference to external objects but the objects themselves are entirely independent of human knowing apart from appearances preconditioned by epistemic conditions. Kant's transcendental idealism held that what can be known are the appearances and never the things in themselves, never those objects external to a knowing subject and never independent of the preconditioned epistemic cognitive conditions of the knower.

However, as Robert Paul Wolff indicates, beyond this distinction between the sensible and intelligible, Kant "moved steadily toward the view that all philosophical investigations of the nature of being or metaphysics must be transformed into a critical analysis of the nature of knowing."[53] Within the 1770 *Dissertation* Kant makes a distinction between two sources of concepts: sensibility and intelligibility. Sensibility is the source of empirical concepts such as concepts of cats and dogs, or houses while intelligence is the source of pure concepts, such as substance, cause, possibility, existence, and necessity.[54] Thus for Kant concepts such as substance, cause, necessity, and existence were to be analyzed in terms of their source in pure concepts and the nature of knowing. This would later become the *Critical Transcendental Philosophy* of Kant where the understanding itself was conceptualized in terms of transcendental idealism in the synthetic unity of apperception of consciousness, and concepts of metaphysics such as causality, necessity, universality, existence, etc. would be subsumed under the rules or forms of sensibility called the schema in the *Critique*.

---

things in themselves." (*CPR*, A490-91/B518-19). A case in point is where Aquinas explains that truth is in judgment. Since the senses have judgment in a limited sense, truth is in the senses only in a limited sense. Truth is properly found in the intellect as it apprehends and judges that there is conformity between what is thought about an external object and the external object itself: "Truth is both in intellect and in sense, but not in the same way… Truth follows the operation of the intellect inasmuch as it belongs to the intellect to judge about a thing as it is … Truth is in sense also as a consequence of its act, for sense judges of things as they are… sense judges truly about things" (*De Veritate* I, q. 1, a. 9, body). In contrast, Kant in the *Critique of Pure Reason* indicates that transcendental idealism holds that "all objects of experience are appearances, mere representations, having no independent existence outside our thoughts" (*CPR*, A490-91/B518-19).

53 Robert Paul Wolff, *Kant's Theory of Mental Activity*, (Gloucester, Mass: Peter Smith, 1973), 13.
54 Ibid., 15.

Kant retained after Wolffian logic, scholastic notions where through abstraction of concepts such as "horse," "dog," and "man" one can form a concept of "mammal" allowing for logical beings of reason to be formed, but understood in terms of extension rather than comprehension.[55] Thus Kant retained logical notions such as genus and species but Kant drew from the *via moderna* through Wolffian logic the conception of extension rather than comprehension as the basis for identifying the essential nature of a given genus contrary to the *via antiqua*.[56]

The reason for this distinction between empirical and pure concepts and between sensibility and intelligence was intended to refute the Leibnizean view that sensible representations are less clear than intellectual representations. Leibniz held that there is little difference between sensibility and intelligence except that sensibility is simply more obscure in its presentation of the same (noumenal) existence that the intelligence more adequately represents.[57] Kant in opposition held that sensibility may in fact be very distinct while the intellect might be confused, thus confused representation was an issue of the intelligence and not of sensibility striking at the heart of the rationalist tradition. This is the case because for the rationalist following the mathematical axiomatic model, what was most certain and clear was the intelligible while the senses and perception were less reliable.

Following the *Meditations and Discourse on Method*, Descartes had maintained that the quality of a concept is the measure of the reality of some external object represented by that concept. However, since Descartes had no clear and distinct idea of material objects, Descartes doubted their existence. But his clarity of ideas about himself gave

---

55 Ibid.
56 Wolf seems to understand apprehension and judgment in terms of extension rather than comprehension where the universal is simply the common representation of many instances, not the essence of things. This illustrates the possible nominalism influencing Kant's conception of apprehension and judgment that continues in both the *Dissertation* and in the *Critique*, but Kant's epistemic shift was radically different from that of the late medieval nominalists since Kant reversed the order of knowability going beyond the realism associated with universals that nominalism had rejected.
57 Robert Paul Wolff, *Kant's Theory of Mental Activity*, (Gloucester, Mass: Peter Smith, 1973), 16.

assurance of his own existence. Thus one can either assert that clear and distinct ideas exist by which one is able to claim that objects of our knowledge are independent realities; or one must deny that clear and distinct ideas are possible, *thus retreating into agnosticism about the existence of objects.*[58] Kant maintained that the former were the continental rationalists whom he would call Dogmatists in the *Critique*, and the latter were the British empiricists called the Skeptics.[59]

Kant's solution to the rationalist and empiricist problem introduced by the Cartesian dualism between intelligibility and sensibility, held that it is not the internal clarity of a concept but rather the nature of its origin that was at issue. Kant failed to address the fundamental problem

---

58 Ibid.

59 Aquinas affirms the fundamental reliability of both sensibility and intelligibility in knowing what a thing is, but affirms that conformity or truth and error occurs at the level of composition and division in judgment where a synthesis between a subject and its *predicaments* takes place: "the sense is indeed true in regard to a given thing, as is also the intellect in knowing what a thing is; but it does not thereby know or affirm truth. ... Yet the perfection of the intellect is truth as known. Properly speaking, therefore, truth resides in the intellect composing and dividing, and not in the sense or in the intellect knowing what a thing is" (*ST* I, q. 16, a. 2). Again, Aquinas explains, "truth is found in the intellect according as it apprehends a thing as it is; and in things according as they have being conformable to an intellect" (*ST* I q. 16, a. 5). And once again, "truth resides, in its primary aspect, in the intellect. Now since everything is true according as it has the form proper to its nature, the intellect, in so far as it is knowing, must be true, so far as it has the likeness of the thing known, this being its form, as knowing. For this reason truth is defined by the conformity of intellect and thing; and hence to know this conformity is to know truth" (*ST* I q. 16, a. 2). All citations of *ST* are from *Summa Theologiae,* trans. by English Dominicans (London: Burns, Oates, and Washbourne 1912-36; repr. New York: Benziger 1947-48; repr. New York: Christian Classics, 1981) unless otherwise noted.

Including the *conclusio* Aquinas concludes in question 16 of the *Prima Pars*: "verum est tantumin intellectu componente, vel dividente; non autem in sensu, vel intellectu quid est rei concipiente. Respondeo dicendum quod verum, sicut dictum est (art. praec.), secundum sui primam rationem est in intellectu. Cum autem omnis res sit vera secundum quod habet propriam formam naturae suae, necesse est quod intellectus, inquantum est cognoscens, sit verus, inquantum habet similitudinem rei cognitae, quae est forma ejus, inquantum est cognoscens. Et propter hoe per conformitatem intellectus et rei veritas definitur. Unde conformitatem istam cognoscere, est cognoscere veritatem" (*ST* I, q. 16, a. 5; *ST*, I, q. 16, a. 5, *conclusio* and *respondeo* par. 1, ed. Duodecima, vol. I, 144). For Aquinas there is no falsity in the apprehension of the essence or quiddity of a thing, and truth is not in knowing what a thing is because the essence or quiddity of a thing is apprehended in the first act of the intellect and is not apprehended in composition or division. Instead truth is found in the second act as the intellect composes and divides a thing as it is, and thus truth is the conformity between the form of the thing and the likeness of the thing known which is the intentional form of that thing in the intellect.

In contrast to the early moderns who after Berkeley found no reliable basis for primary or secondary qualities, Aquinas maintained that proper sensibles (secondary qualities) such as color, flavor, odor, etc. are reliably known since they are directly associated with the sense organs as color is to sight and flavor is to taste, but common sensibles requiring more than one sense can be mistakenly judged as is the case with incidental sensibles that are not associated directly with any sense.

of the reliability of sensation and its relation to the intellect and rather emphasized the nature of the origin of the internal clarity of concepts. Kant argued that ideas are in fact the measure of the reality of the object rather than the external object being the measure of ideas. Kant thus deviated significantly from the *via antiqua* of Aquinas where existence external to a knowing subject is the measure of the human intellect rather than the human intellect being the measure of existence, and where the divine intellect (divine ideas) rather than the human intellect is the measure of existence.[60]

Aquinas explains that truth in the intellect is found according to its conformity with things whose notes the intellect apprehends through simple apprehension as the essence or form of the thing. The truth in things is found according as things imitate the divine intellect, and the divine intellect is their measure according as they by their very nature bring about a true apprehension of themselves in the human intellect where the human intellect is measured by things. The form of a thing existing external to a knowing subject imitates the art of the divine intellect, and by this same form, the form is such that it can bring about a true apprehension of itself and can be judged truly in the human intellect.[61] Thus for Aquinas and the *via antiqua* our intellect

---

60 Allison argues that any transcendental realist project is essentially a theocentric model but Kant's emphasis is upon an anthropocentric model. Cf. Henry E. Allison, *Kant's Transcendental Idealism* (New Haven: Yale University Press, 1983), 19-25. A theocentric model in general entails human knowledge being analyzed in terms of its conformity, or lack thereof, to a standard of cognition set by an "absolute" or "infinite intelligence." Leibniz following Augustine and Malebranche, held that the divine intellect is where one finds "the pattern of the ideas and truths which are engraved in our souls" (cf. G. W. Leibniz, *New Essays On Human Understanding*, trans. Peter Remnant and Jonathan Bennett, book 4, chap. 2, §14, p. 447). Aquinas following Augustine maintained that divine ideas are the measure of things (*De Veritate* I, q. 3, a. 1).

61 Aquinas explains the relationship: "among created things truth is found both in things and in intellect. In the intellect it is found according to the conformity which the intellect has with the things whose notions it has. In things it is found according as they imitate the divine intellect, which is their measure--as art is the measure of all products of art--and also in another way, according as they can by their very nature bring about a true apprehension of themselves in the human intellect, which, as is said in the Metaphysics (Aristotle, *Metaph.*, {I}, 1 (1053a 31)), is measured by things. By its form a thing existing outside the soul imitates the art of the divine intellect; and, by the same form, it is such that it can bring about a true apprehension in the human intellect. Through this form, moreover, each and every thing has its act of existing. Consequently, the truth of existing things includes their entity in its intelligible character, adding to this a relation of conformity to the human or divine intellect. But negations or privations existing outside the soul do not have any form by which they can imitate the model of divine art or introduce a

is measured by natural things, and measures only artifacts, not natural things.[62] Aquinas is very explicit on this particular point:

> The true, therefore, is found secondarily in things and primarily in intellect.... Since the speculative intellect is receptive in regard to things, it is, in a certain sense, moved by things and consequently measured by them. It is clear, therefore, that, as is said in the Metaphysics (Aristotle, *Metaph.*, {E}, 4 (1027b 26). {I}, 1 (1053a 33); {I}, 6 (1057a 9-13)), natural things from which our intellect gets its scientific knowledge measure our intellect. Yet these things are themselves measured by the divine intellect, in which are all created things--just as all works of art find their origin in the intellect of an artist. The divine intellect, therefore, measures and is not measured; a natural thing both measures and is measured; *but our intellect is measured, and measures only artifacts, not natural things.*[63]

For Aquinas the quality of the external object signified by a given concept was not in question although it was understood that perception has its limitations, but truth and falsity were found properly speaking in the intellect and not in the senses *per se* nor in apprehending the quiddity or essence of a thing.[64] Aquinas explains:

---

knowledge of themselves into the human intellect. The fact that they are conformed to intellect is due to the intellect, which apprehends their intelligible notes" (*De Veritate* I, q. 1, a. 8, body).

62 It is on this very point that "transcendental thomism" either stands or falls. If Aquinas held that objects external to a knowing subject measure the intellect, then "transcendental thomism" cannot embrace transcendental idealism by definition and remain epistemological Thomists.

63 *De Veritate* I, q. 1, a. 2, body.

64 Aristotle in explaining the relation between the special, incidental, and common sensibles explains: "Perception of the special objects of sense is never in error or admits the least possible amount of falsehood. Next comes perception that what is incidental to the objects of perception is incidental to them: in this case certainly we may be deceived; for while the perception that there is white before us cannot be false, the perception that what is white is this or that may be false. Third comes the perception of the common attributes which accompany the incidental objects to which the special sensibles attach (I mean e.g. of movement and magnitude); it is in respect of these that the greatest amount of sense-illusion is possible" (*De Anima* III, 428b17-428b26). All citations of *De Anima* from Aristotle, *The Complete Works of Aristotle The Revised Oxford Translation* Vol I. Edited. by Jonathan Barnes, Vol. II, *De Anima* translation by W. D. Ross and revised by J. O. Urmson. New Jersey: Princeton University Press, 1984. Likewise, Aquinas writes: "What is commonly asserted by all, cannot be wholly false. For a false opinion betrays a certain weakness of intellect, just as a false judgment about a proper object of sense

The Philosopher (*De Anima* iii, 6) compares intellect with sense on this point. For sense is not deceived in its proper object, as sight in regard to color; has accidentally through some hindrance occurring to the sensile organ---for example, the taste of a fever-stricken person judges a sweet thing to be bitter... Sense, however, may be deceived as regards common sensible objects, as size or figure; when, for example, it judges the sun to be only a foot in diameter, whereas in reality it exceeds the earth in size. Much more is sense deceived concerning accidental sensible objects, as when it judges that vinegar is honey by reason of the color being the same. The reason of this is evident; for every faculty, as such, is "per se" directed to its proper object; and things of this kind are always the same. Hence, as long as the faculty exists, its judgment concerning its own proper object does not fail. Now the proper object of the intellect is the "quiddity" of a material thing; and hence, properly speaking, the intellect is not at fault concerning this quiddity; whereas it may go astray as regards the surroundings of the thing in its essence or quiddity, in referring one thing to another, as regards composition or division, or also in the process of reasoning. Therefore, also in regard to those propositions, which are understood, the intellect cannot err, as in the case of first principles from which arises infallible truth in the certitude of scientific conclusions.

The intellect, however, may be accidentally deceived in the quiddity of composite things, not by the defect of its organ, for the intellect is a faculty that is independent of an organ; but on the part of the composition affecting the definition,

---

results from a weakness in the sense. But defects are accidental, because they are outside the intention of nature. What is accidental, however, cannot exist always and in all beings; thus the judgment which every taste registers about flavors cannot be false." (*On the Eternity of the World* II, Ch 34). All citations of *On the Eternity of the World* are from *On the Eternity of the World*, trans. by Cyril Vollert (Milwaukee: Marquette University Press, 1964) unless otherwise noted.

when, for instance, the definition of a thing is false in relation to something else, as the definition of a circle applied to a triangle; or when a definition is false in itself as involving the composition of things incompatible; as, for instance, to describe anything as "a rational winged animal." Hence as regards simple objects not subject to composite definitions we cannot be deceived unless, indeed, we understand nothing whatever about them, as is said Metaph. ix, Did. viii, 10 (body emphasis added).[65]

If the operation of the intellect is traced to a thing as its cause, then in composite substances the combination of matter and form, or of subject and accident, the thing itself serves as the foundation and cause of the truth in the combination which the intellect makes and expresses in words for Aquinas.[66] Of course, this applies when speaking of some being in nature, but when speaking of intentional beings of reason the intentional being itself rather than *res naturae* serves as the foundation and cause of the truth.

Aquinas does not expect numerical identity between an external object and the object as known. There is identity in the sense of conformity between a given subject and its predicate. The intentional stone is a likeness of the external stone and not according to the mode of being which that stone has in reality as it exists external to a knowing subject, but according to the mode of being of the knowing subject. Aquinas explains, "For in the soul there is not the form by which a stone exists but only a likeness of that form. Hence, what is necessary is not that the form be individuated, but that it be a likeness of the form that is individuated."[67]

For the empiricists and rationalists, rather than having a concern with the realism or nominalism associated with *quiddities*, since nominalism was assumed by the early modern period, concepts

---

65 *ST* I, q. 85, a. 6.
66 *In Metaph.*, *IX* lect. 11, no. 1898.
67 *De Veritate* VIII, a. 11, ad2-4.

themselves determined the clarity or lack of clarity and thus it became an issue of the internal clarity of ideas or concepts that were at issue for the early moderns. Further, early modern philosophy ignored the role of incidental sensibility in causality limiting perception to primary and secondary qualities that became increasingly unreliable thus early modern philosophy was unable to explain any sensible-cognitive connexion between a given cause and its effect. Therefore due to these early modern developments, early moderns were no longer able to demonstrate God's existence by his effects. Galileo would write in the Assayer:

> I think, therefore, that these tastes, odors, colors, etc, so far as their objective existence is concerned, are nothing but mere names for something which resides exclusively in our sensitive body, so that if the perceiving creatures were removed, all of these qualities would be annihilated and abolished from existence… I cannot believe that there exists in external bodies anything, other than their size shape, or motion (slow or rapid), which could excite in us our tastes, sounds, and odors. And indeed I should judge that, if ears, tongues, and noses be taken away, the number, shape, and motion of bodies would remain, but not their tastes, sounds, and odors. The latter, external to the living creatures, I believe to be nothing but mere names.[68]

The rationalists and empiricists inherited Galileo's and Descartes skepticism toward secondary qualities also known as proper sensibles by Aquinas as well as Galileo's nominalism. That which could be measured such as motion and extension were held to exist external to a perceiving subject, but secondary qualities such as flavor, colors, and odors were not held to exist independent of one sensing or perceiving such qualities. Therefore, the early moderns came to hold that such secondary qualities were merely names associated with perception and

---

68 Galileo's *Il Saggitore* (The Assayer), which appeared originally in 1623. Translation by A. C. Danto, from *Introduction to Contemporary Civilization in the West* Vol. I, 2nd ed. (New York: Columbia University Press, 1954), 719-724.

were therefore merely ideas. Thus both the rationalist and empiricist traditions asserted clear and distinct ideas were at issue: "that the objects of our knowledge are independent realities as in the case of the rationalists; or one denies that we can have clear and distinct ideas, and thereby retreats into agnosticism about the existence of objects as in the case of the empiricists."[69] To avoid these two alternatives, Kant attempted to reject the premise on which these alternatives were based. Hume had argued that reason cannot infer a cause from a given effect nor can any *connexion* between the two be inferred. Hume further explained that any *a priori* relation between a cause and effect is arbitrary:

> Nor does any man imagine that the explosion of gunpowder, or the attraction of a loadstone, could ever be discovered by arguments *a priori*.... In a word, then, every effect is a distinct event from its cause. It could not, therefore, be discovered in the cause, and the first invention or conception of it, *a priori*, must be entirely arbitrary.... When we reason *a priori*, and consider merely any object or cause as it appears to the mind, independent of all observation, it never could suggest to us the notion of all observation, it never could suggest to us the notion of any distinct object, such as its effect; much less show us the inseparable and inviolable connexion between them."[70]

Hume went further however and argued that one constantly conjoins together various events and objects and from this experience infers the existence of one object from the appearance of another, but it is not reason or understanding that draws this inference, it is custom or habit, and ultimately belief which is felt by the mind enforcing custom in the mind and rendering them the governing principles of our actions.[71] By this a given concept arises from a *customary* conjunction

---

69 Robert Paul Wolff, *Kant's Theory of Mental Activity*, (Gloucester, Mass: Peter Smith, 1973), 16.
70 David Hume, *An Enquiry Concerning Human Understanding* IV, Part I in *Hume: Theory of Knowledge*, ed. D. C. Yalden-Thomson, (New York: Thomas Nelson and Sons, 1951), 27-29, 31.
71 Hume explains: "Yet he has not, by all his experience, acquired any idea or knowledge of the secret

formed between an object of experience and something present in memory or the senses. From the impressions of belief Hume argued that "principles of connexion or association" are formed and can be reduced to three: resemblance, contiguity, and causation which unite thoughts together.[72] But Hume argues that "in these phenomenon, the belief of the correlative object is always presupposed; without which the relation could have no effect... Contiguity to home can never excite our ideas of home, unless we believe that it really exists" and this independent of reason.[73]

---

power by which the one object produces the other; nor is it by the process of reasoning he is engaged to draw this inference. But still he finds himself determined to draw it: and though he should be convinced that his understanding has no part in the operation, he would nevertheless continue in the same course of thinking. There is some other principle with determines him to form such a conclusion. This principle is custom or habit. For whatever the repetition of any particular act or operation produces a propensity to renew the same act or operation, without being impelled by any reasoning or process of the understanding, we always say that this propensity is the effect of custom.... After the constant conjunction of two objects—heat and flame, for instance, weight and solidity—we are determined by custom alone to expect the one from the appearance of the other. This hypothesis seems even the only one which explains the difficulty why we draw, from a thousand instances, an inference which we are not able to draw from one instance that is, in no respect, different from them. Reason is incapable of any such variation... no man, having seen only one body move after being impelled by another, could infer that every other body will move after a like impulse. All inferences from experience, therefore, are effects of custom, not of reasoning. Custom, then, is the great guide of human life. It is that principle alone which renders our experience useful to us, and makes us expect, for the future, a similar train of events with those which have appeared in the past. Without the influence of custom, we should be entirely ignorant of every matter of fact beyond what is immediately present to the memory and senses.... What, then, is the conclusion of the whole matter?... All belief of matter of fact or real existence is derived merely from some object, present to the memory or senses, and a customary conjunction between that and some other object. Or in other words, having found, in many instances, that any two kinds of objects—flame and heat, snow and cold—have always been conjoined together; if flame or snow be presented anew to the senses, the mind is carried by custom to expect heat or cold, and to believe that such a quality does exist, and will discover itself upon a nearer approach. It is an operation of the soul, when we are so situated, ... All these operations are a species of natural instincts, which no reasoning or process of the thought and understanding is able either to produce or to prevent...belief is nothing but a conception more intense and steady than what attends the mere fictions of the imagination, and that this manner of conception arises from a customary conjunction of the object with something present to the memory or senses. " David Hume, *An Enquiry Concerning Human Understanding* IV, Part I in *Hume: Theory of Knowledge*, ed. D. C. Yalden-Thomson (New York: Thomas Nelson and Sons, 1951), 42-47.

72 David Hume, *An Enquiry Concerning Human Understanding* IV, Part II in *Hume: Theory of Knowledge*, ed. D. C. Yalden-Thomson (New York: Thomas Nelson and Sons, 1951), 51.

73 Ibid., 54. Hume immediately continues to argue even in thought any such relationships of connexion is from experience and not from reason: "I assert that this belief, where it reaches beyond the memory or senses, is of a similar nature, and arises from similar causes, with the transition of thought and vivacity of conception here explained. When I throw a piece of dry wood into a fire, my mind is immediately carried to conceive that it augments, not extinguishes, the flame. This transition of thought from the cause to the effect proceeds not from reason. It derives its origin altogether from custom and experience. And as it first begins from an object present to the senses, it renders the idea

Hume then argued that "beyond the constant conjunction of similar objects, and the consequent inference from one to the other, we have no notion of any necessity or connexion."[74] Kant realized that Hume had argued that causality and necessity are not *a priori* and are independent of reason, and Kant realized that Hume argued that causality and necessity are inferences from custom and were simply impressions of belief from which the mind infers necessity and causation. Necessity and causation arise entirely from the uniformity observed in nature, where similar objects are constantly conjoined together, and the mind is determined by custom to infer a causal or necessary connection of one object to another by the appearance of the latter object. Kant supposed that such a position would mean that necessity and causality were entirely arbitrary and this would essentially be the end of both science and metaphysics. Thus both the rationalist and empiricist traditions asserted clear and distinct ideas were at issue. To avoid these two alternatives, Kant attempted to reject the premise on which these alternatives were based.

Awakened from his Wolffian dogmatic or rationalist slumber, for Kant it was no longer an issue of the internal clarity of concepts or the means by which concepts apprehend or represent independent realities,

---

or conception of flame more strong and lively than any loose, floating reverie of the imagination. That idea arises immediately. The thought moves instantly towards it, and conveys to it all that force of conception which is derived from the impression present to the senses... what is there in this whole matter to cause such a strong conception, except only a present object and a customary transition to the idea of another object, which we have been accustomed to conjoin with the former? This is the whole operation of the mind, in all our conclusions concerning matter of fact and experience;..." (Ibid., 54-55).

74 David Hume, *An Enquiry Concerning Human Understanding* VIII, Part I in *Hume: Theory of Knowledge*, ed. D. C. Yalden-Thomson (New York: Thomas Nelson and Sons, 1951), 85. The entire context follows: "If every object was entirely new, without any similitude to whatever had been seen before, we should never, in that case, have attained the least idea of necessity, or of a connexion among these objects. We might say, upon such a supposition, that one object or event has followed another; not that one was produced by the other. The relation of cause and effect must be utterly unknown to mankind. Inference and reasoning concerning the operations of nature would from that moment be at an end; and the memory and senses remain the only canals by which the knowledge of any real existence could possibly have access to the mind. Our idea, therefore, of necessity and causation arises entirely from the uniformity observable in the operations of nature, where similar objects are constantly conjoined together, and the mind is determined by custom to infer the one from the appearance of the other. These two circumstances form the whole of that necessity which we ascribe to matter. Beyond the constant conjunction of similar objects, and the consequent *inference* from one to the other, we have no notion of any necessity or connexion" (Ibid., 84-85; emphasis added).

but rather it became an issue of the nature and origin of concepts becoming the measure of reality of its objects rather then the objects being the measure of the reality of concepts. This was the fundamental shift between the Kantian transcendental turn and that of the various epistemologies prior to the Kantian Copernican revolution.[75]

This shift was itself driven by Hume's critique of reason and Hume's argument that only experience and custom give us any notion of causality or necessity taken as similarity between independent events, and from experience reason itself cannot infer a causal connection between a given cause and its effect since any such inference is simply a belief or validity formed by like impressions and custom. However, in developing an alternative, Kant continued to embrace the fundamental nominalism of early modern philosophy as expressed by Galileo but

---

75 Leo Sweeney explains: "Immanuel Kant is directly opposed to Aquinas, since for the latter the known, not the knower, is the sole content-determining cause of knowledge, whereas for the former the knower through his/her twelve categories and the *a priori* forms of space and time determines the intellectual content of the known, which is thus solely the "phenomena" constructed by the mind and not the "noumena," which remains unknown. Consequently, Kantianism so taken seems so opposed to Thomism that "transcendental Thomism" is almost a contradiction in terms." Sweeney then indicates that Karl Rahner, Bernard Lonergan, and Emerich Coreth are in opposition to Thomism according to Leslie Dewart who argues that any form of Kantianism is in opposition to Thomism. However, Sweeney warns it is not so simple since those who follow for example Joseph Marechal's school (1878-1944) are not so "patently Kantian." Instead, their position often mixes Kant with insights from various philosophical schools resulting in a perspective radically different from Thomas Aquinas. The Belgian Jesuit maintained that the Kantian *Critiques* had to be taken seriously and thus Marechal reintroduced in transcendental Thomism the epistemological problem raised by Kant, but rather than addressing the problem of the epistemological turn, the Marechalian school gave a reinterpretation of basic Thomistic insights in light of the Kantian crises. Cf. Leo Sweeney, "Must Thomism Become Kantian" in *Thomistic Papers VI* (Texas: The Center for Thomistic Studies, 1994), 172-173, 182-183, 194; emphasis added; also see L. Dewart, *The Future of Belief* (New York: Herder and Herder, 1969), Appendix 2: "On Transcendental Thomism," 499-522; also Lawrence K. Shook, *Etienne Gilson* (Toronto: Pontifical Institute of Mediaeval Studies, 1984), 171.

Likewise Henle explains, "In accepting the Kantian methodology, Transcendental Thomists implicitly accepted Kant's formulation of the problem of human knowledge. Kant's formulation in turn rested on Hume's conclusion that no necessity, no universality and no intelligibility can be found in experience. The Humean presupposition infects the whole of the Critique of Pure Reason and consideration of it reveals the transcendental method as a tour-de-force to solve a false problem.

The Humean presupposition is why the Transcendental Thomists insist so strongly that there is no intuition of being in experience. It is also why Joseph Donceel wrote: 'That is why metaphysics is *a priori*, virtually inborn in us, not derived from experience, … Yet we would never know any metaphysics if we had no sense knowledge… Hence metaphysics is *virtually* inborn in us. We become aware of it only *in* and *through* sense knowledge, although it does not come *from* sense knowledge'" (Robert J. Henle, "Apropos of McCool" in *Thomistic Papers VI* (Texas: The Center for Thomistic Studies, 1994), 144-145).

Kant shifted the focus from sensibility or empiricism to consciousness or rationalism driving a further wedge between sensibility and intelligibility.

In the case of the *via moderna*, Occam's nominalism suggested that the object signified by the concept itself had only nominal existence as that which refers to a given external object as far as similarity between external objects are concerned, and therefore for Occam the object of the concept or that which the concept signified was no longer the thing itself taken as an intelligible form or similitude.[76] Consequently, there arose a concern in early modern philosophy for the internal clarity of ideas since such ideas do not represent the thing itself but our conception of that thing.

However, Kant wished to maintain some form of internal clarity of ideas, but Kant continued the disparity between sensibility and

---

76 Maritain notes that for nominalism, the concept has no reality other than the individuals, which the concept represents, and therefore it derives its character from extension rather than comprehension. That is for nominalism, the concept finds its universality through it applicability to a greater or lesser group of individuals rather than from the essence, nature, or quiddity presented immediately to the mind. The nominalist theory of the *via moderna* abstracts the extensionality of a given concept from external objects in opposition to the Kantian noetic, but similar to the Kantian noetic concepts only contain what one puts into them. Thus the *via moderna* would be considered essentially a transcendental realist account by Kant rather than a transcendental idealist epistemology. Since concepts, regardless of content, are abstracted from external objects and are measured by the external objects, both the *via moderna* and the *via antiqua* share in Kant's critique against transcendental realism. In the *via moderna* account "we do not attain through our concepts essences or natures which are what they are in themselves, independent of the manner in which we apprehend them" as would be the case in the *via antiqua* account of Aquinas. Maritain thus concludes that this suggests that in the *via moderna* theory, "a concept may only be understood in a subjective sense; it is merely a group of notes which we have explicitly collected and which, given our present state of knowledge, constitute the concept *for us*." Although it may be the case that such concepts are subjective in the *via moderna* theory of abstraction, nominalism did not hold that such concepts were determined by *a priori* epistemic conditions. In contrast, however to the *via moderna* account, for the *via antiqua*, the essence or comprehension is something real, and the concept is *essentially and originally characterized by its comprehension*. Comprehension is simply the sum of the constitutive notes of the nature or essence as presented to the intellect in abstraction via the first operation of the intellect. Thus comprehension in contrast to the *via moderna*, is to be understood in an objective sense where the "comprehension of the concept is the sum of the notes that constitute the concept *in itself*: first and foremost, notes that constitute the very *essence* that is presented by this concept (e.g. rationality and animality in the concept man), secondarily, the notes that are necessarily derived from this essence and fundamentally contained in its constitutive notes (properties, such as "risibility," "gifted with articulate speech," etc., which are brought to light by the reasoning, and are contained virtually in the concept, although as yet unknown to us)." In the *via antiqua*, in contrast to the *via moderna*, the extension of the concept is a property following upon abstraction, and extension presupposes comprehension where the concept is universal *only because it reveals the necessary condition of some essence*. Cf. Jacques Maritan, *Formal Logic*, (New York: Sheed & Ward, 1946), 22-23.

intelligibility in attempting to do so. As Robert Wolff explains, Kant maintained that "representations derived from sensibility reveal only appearance, no matter to what pitch of systematic order and distinctness they are brought by logic. Representations derived from intelligence, on the other hand, reveal things as they are in themselves, even if only dimly and with confusion."[77] Kant does in fact give validity to early modern metaphysics but denies that we have sensitive knowledge of substance, causality, possibility, existence, and necessity independent of synthetic *a priori* epistemic conditions. Kant also grants validity to mathematics in geometry and physics, which were essential to the new Newtonian inductive experimental mathematical science, but Kant restricts their scope of application to appearances (i.e., phenomena).[78]

Again as early as the 1770 publication of the *Dissertation* Kant had determined that there are two sensible epistemic conditions, space and time, which are "schemata and conditions of all human knowledge that is sensitive."[79] However, in contrast to the Aristotelian categories of place, location, and time, the Kantian notions of space and time were abstracted but not from sensing objects since sensation can only give the matter, not the form, of human apprehension. Instead for Kant space and time were abstracted from the action of the mind *in co-ordinating its sensa according to unchanging laws.*[80] Note again Kant assumes that sensation can only give the matter and never the form *of human apprehension*, but the *via antiqua* would argue that it is the form that is apprehended in simple apprehension as all material conditions are removed from sensible objects of knowledge and what remains is generality or the universal.

Thus sensation does in fact give matter as Kant held, but in opposition to Kant, the form is abstracted in simple apprehension from sensation entirely independent of the "*a priori* rules in virtue of which intuitions come to represent an objective world of causally interacting

77 Robert Paul Wolff, *Kant's Theory of Mental Activity* (Gloucester, Mass: Peter Smith, 1973), 16.
78 Ibid., 17.
79 Ibid.
80 Ibid.

substances standing in spatio-temporal relations to one another" and independent of the "empirical concepts in virtue of which intuitions come to represent objects as having determinate empirical features."[81]

For Aquinas in opposition to Kant, the imagination simply does not synthesize sensible forms according to *a priori* or empirical schemata since any such activity would be that of the intellect rather than an activity of the imagination and even then it is not the schemata that determines what can be known of existence but existence that determines what can be immediately known of the categories. Further, for Aquinas like Aristotle sensible forms are not given order or unity by *a priori* rules or by empirical schemata, but rather such unity and order is apprehended from existence itself for Aquinas. For example, rather than an empirical schemata defining the color red in this or that intuition of a red apple, the red apple itself determines the synthetic unity within the imagination of a red apple as apprehended from this or that red apple.

Robert Wolff seems to note a distinctive difference between the characteristics of the forms of sensibility given in the *Dissertation* and the later *Transcendental Aesthetic*. In the *Dissertation*, the forms of sensibility of space and time are given the function of coordinating the sensible material according to these forms of sensibility. Thus physics and geometry are derived from the sensible forms of time and space, but according to Wolff, in the *Critique* space and time are passive forms of intuition, which "merely presents a manifold or variety of sensuous material to the synthesizing understanding."[82]

81 For Kant experience consists in the reception of representations of sensibility as well as synthesis of intuitions, that is of singular representations. These intuitions are structured or synthesized by the imagination according to rules, some of which are *a priori* and others are empirical rules both of which are called schemata. The *a priori* rules allow for "intuitions to represent an objective world of causally interacting substances standing in spatio-temporal relations to one another. The *a priori* rules are the schemata in virtue of which the pure concepts of understanding are applicable to experience. But there are also rules that correspond to our empirical concepts, in virtue of which intuitions come to represent objects as having determinate empirical features" (Ginsborg, 2002, 4). For example, empirical rules allow a given object of possible experience to have the quality of green or blue. It is the imagination that synthesizes the various intuitions according to both *a priori* and empirical schemata. See Hannah Ginsborg, "Thinking the Particular as Contained under the Universal," Department of Philosophy, U.C. Berkeley Draft (November, 2002): 4.
82 Ibid., 18.

Thus Kant had not arrived at the notion that all activity must be attributed to intelligence in the *Dissertation*; however, the Aesthetic does embody the *Dissertation's* position on space and time, virtually unchanged even to the arguments used in the Metaphysical and Transcendental Expositions. But the distinctive difference is that in the *Dissertation* Kant held that "sensibility itself contains the laws for the coordination of a sensuous manifold."[83] But in the *Critique*, Kant assigns the process of synthesis to understanding, and therefore all knowledge can only be explained in reference to the contribution of intelligence. This causes Kant to reverse himself in the *Axioms of Intuition* and Kant grounds mathematics, as a knowledge of appearances, on the *categories of quantity* in the *Critique*. Thus, Kant's later development following 1770 was the introduction of the categories and his assignment of the process of synthesis to understanding.

In addition, Kant argued in the *Dissertation* that if all properties of space are borrowed only from external relations through experience, geometrical axioms could never possess universality or any basis of certainty, but only that comparative universality acquired through induction and holds only so widely as it is observed.[84] Thus Kant begins to turn his investigation to the universal and necessary grounding of mathematics and in the years following 1770 began to realize that concepts such as substance, causality, possibility, existence, and necessity needed a similar grounding as that based upon the forms of space and time. According to Robert Wolff, "the Critique is in large measure the outcome of this realization."[85]

Thus to address the transcendental idealism introduced by the *Critique* one must first take into account Kant's need to ground universality and necessity for mathematics, substance, causality, and so forth by accounting for the possibility of universality and necessity.[86]

---

83 Ibid.
84 Ibid., 19.
85 Ibid.
86 For Aquinas universality was attained by abstraction from sensible conditions and necessity could in fact be abstracted from sensibility as in the case of natural necessary modality as well as in discursive reason from logical syllogistic necessity where a necessary conclusion is deduced from principles arrived

In the 1770 Dissertation Kant had not fully realized the importance of the problem of causation and held that pure concepts are abstracted by attention to its actions on the occasion of experience from laws inborn in the mind, but Kant later realized that Hume's critique made this rationalist view indefensible.[87]

Kant makes another turn described in the Dissertation under the section *On the Method of Dealing with the Sensitive and the Intellectual in Metaphysics*. It is here that Kant asserts the origin and nature of error in metaphysics in allowing the *principles proper to sensitive apprehension to pass from their boundaries in sensibility to meddling with the intellectual*. At this stage it would appear that Kant was following Christian Wolff who consistently maintained with the scholastics that there are three operation of the intellect, with the second operation being that of judgment where subject/predicate propositions are formed.[88] Although Kant agrees that the subject is unthinkable apart from the condition of the predicate, "the predicate is a condition without which the subject is asserted to be unthinkable," yet Kant held that if the predicate is a

---

at inductively. Likewise Aristotle explains: "Now induction is of first principles and of the universal and deduction proceeds from universals. There are therefore principles from which deduction proceeds, which are not reached by deduction; it is therefore by induction that they are acquired" (*Nich. Eth.* Bk6 1139b19-1139b35).

87 Robert Paul Wolff, *Kant's Theory of Mental Activity*, (Gloucester, Mass: Peter Smith, 1973), 19-20.

88 Christian Wolff explained: "there are three operations of the intellect, namely, Notion with simple apprehension, Judgment, and Discourse. The first operation of the intellect is the representation of many things, one by one, in a single thing.... Judgment comes from use and exercise in joining many things together." Note it seems that Wolff understands apprehension and judgment in terms of extension rather than comprehension where the universal is simply the representation of many things. This illustrates the possible nominalism influencing Kant's conception of apprehension and judgment. Swedenborg notes that for Wolff "the representation of a thing when considered objectively is called an idea; the representation of things or species in a universal, is called a notion. Notions like perceptions are clear or obscure; distinct or confused. To cognize is to acquire an idea or notion of that thing and cognition is an action of the soul where the faculty of cognizing is that by which we acquire ideas and notions." It seems that Wolff's conception of the universal was a representation of many things or species, taken "one by one, in a single thing" rather than the apprehension of the essence taken in terms of comprehension of a given thing suggesting a nominalist *via moderna* theory of abstraction influencing Kant's conception of universality only in terms of extension. Likewise Gilson was correct to assert that Wolff had detached essence from existence, the idea was understood primarily in terms of extension rather than comprehension and extension. Thus it is likely that Kant's notion of the universal is that of nominalism, but Kant's noetic turn was in opposition to the "transcendental realism" of nominalism. Cf. Emanuel Swedenborg, *Psychologica, Being Notes and Observations on Christian Wolff's Psychologia Empirica*, (Philadelphia: Swedenborg Scientific Association, 1923), 118, 28-36, 14.

concept of sense, it will at most contain a condition with which there is the possibility of sensitive knowledge of a given concept.

For example, given the judgment 'whatever is, is somewhere and somewhen', the predicate is a sensitive concept involving space and time. Therefore, it cannot be considered a universal condition of the subject; instead, it is a condition only of the sensitive apprehension of the subject. However, for the predicate to be considered a universal condition of the subject, the proposition would need to be revised to state: "whatever is, insofar as it is known as an appearance, is somewhere and somewhen." Its predicate thus becomes intellectual and unhindered by the subjective conditions of knowledge associated with sensibility. Hence, the subjects sphere of applicability is universal, including 'what is somewhere and somewhen'. In the *Dissertation*, one finds Kant attempting to preserve metaphysics by separating the sensible conditions of human knowledge from the intellect where the intellect provides the universal and necessary basis for metaphysics.[89] However, in the *Critique* Kant reconciles this difference and attempts to explain how one can maintain the universality and necessity required by metaphysics and mathematics, while holding that synthetic judgments can be both *a priori* and apprehended from sensible conditions of human knowledge.[90]

The commentator Wolff indicates that Kant is the focal point of modern philosophy to which flow the main streams of seventeenth- and eighteenth-century thought, and from whom arise nineteenth-century English and Continental philosophy.[91] The *Critique* was an attempt at a compromise between Continental rationalism and British empiricism. The problem that Kant begins with is how to allow for there to be legitimate claims both by Leibnizians and Newtonians. Kant's distinction between appearance and existence as such allows Kant to save science and make room for metaphysics. However, Kant

---

89 Robert Paul Wolff, *Kant's Theory of Mental Activity*, (Gloucester, Mass: Peter Smith, 1973), 21.
90 See the section titled *The Kantian Synthetic a priori Solution* in the following chapter titled *The Problem of the Synthetic a priori*.
91 Robert Paul Wolff, *Kant's Theory of Mental Activity*, (Gloucester, Mass: Peter Smith, 1973), 21.

sets in motion the destruction of metaphysics by turning from modes of being to ways of knowing. Discovering the necessity of the conditions of the possibility of sensitive knowledge in the *Dissertation*, Kant's critique of the knowledge of things in themselves led metaphysics or being *qua* being to yield its domain to the critique of pure judgment or to the conditions of human knowledge.

The *Dissertation* did not explain how pure concepts can be applicable to noumena, and still be given through the very nature of the intellect. Hence, Kant was left with explaining along with Descartes and Leibniz the question: "how can representations that have their origin in the mind nevertheless give us knowledge of independently existing substances?"[92] Kant mentions in a letter to Marcus Herz dated 21 February 1772 this discovery where Kant explains that it is possible to imagine two kinds of intelligence, the *intellectus archetypus*, on whose intuition the things themselves are grounded, and of an *intellectus ectypus* which derives the data from the sensuous intuition of things.[93]

Robert Wolff quotes the letter to Marcus Herz from Kemp Smith's Commentary. In this letter Kant continues by stating, "But our understanding is not the cause of the object through its representations, nor is the object the cause of its intellectual representation (*in sensu reali*). Hence, the pure concepts of the understanding cannot be abstracted from representations through the senses. But, whilst they have their sources in the nature of the soul, they originate there neither as the result of the action of the object upon it, nor as themselves producing the object."[94] Kant indicates the problem is that pure concepts of the understanding cannot be abstracted from representations through the senses nor can pure concepts produce the objects because the understanding cannot be the cause of objects through its representations, nor can objects be the cause of its intellectual representation. In realizing this, Kant asked, how can representations that originate in the mind give knowledge of independently existing things?

92 Ibid., 22.
93 Immanuel Kant, *Critique of Pure Reason*, trans. by N. Kemp Smith 2nd ed. (London: Macmillan, 1929), 219-220.
94 Ibid.

At this point one can clearly see that Kant's entire epistemological shift is in opposition to the *abstractio* theory held by many if not all forms of transcendental realism. However, it is not because Kant denied that abstraction takes place, but it is because Kant held that such abstraction is of objects of possible experience preconditioned by epistemic rules that govern the *a priori* nature of synthetic judgments. However, Kant was driven to this position by the irreconcilable question of how sensibility and intelligibility can *in unity* form universal and necessary concepts. This entire question itself was however introduced by the rationalist and empiricist debate over the internal clarity of concepts and the reliability of sensibility, and this question in turn was preconditioned by the nominal existence of the concept introduced by the *via moderna*.

Kant accepted that "experience presents us only with particulars and this was derived from the empiricist tradition represented by Locke, Berkeley, and Hume," and in each case they were faced with how to explain how it was "possible to represent general features that were common to a multiplicity of things from sensibility."[95] It was inevitable that Hume's empiricist rejection of the possibility of attaining causality and necessity from reason consequently would lead to Kant's development of the transcendental justification of causality and necessity from pure reason.

Hence, early modern philosophers increasingly began to ignore the principle of abstraction in cognitive psychology and in their critique of sensibility and intelligibility where we find Berkeley's strong polemic against the abstraction theory, but Berkeley still retained an abstraction theory to account for the relationship between sensibility and intelligibility.[96] Hume endorses Berkeley's view but in fact goes beyond it.[97] In maintaining customary or Humean association independent of reason, Hume disconnected reason from experience bringing an end to the possibility of a reasonable reconciliation between intelligibility and

---

95 See Hannah Ginsborg, "Thinking the Particular as Contained under the Universal," Dept. of Philosophy, U.C. Berkeley Draft, (November 2003): 11.

96 Ibid., 12-13.

97 Ibid.

sensibility through the process of abstracting the universal from the particular. Instead as Ginsborg argues, Hume maintained:

> It is a basic psychological fact about us that our associations of ideas follow certain regular patterns, so that, for example, the idea of a particular triangle will naturally call to mind ideas of other triangles in preference, say, to ideas of quadrilaterals or circles or indeed things that are not plane figures at all. It is because of these natural patterns of association that, once the word 'triangle' has been applied to a representative sample of triangles, we will become disposed to apply it to triangles generally;...[98]

Ginsborg argues that Hume's account of empirical generality or abstraction is different from that of Locke and Berkeley, accounting for abstraction or empirical generality by exploiting the generality of custom or disposition.[99] However, for Hume custom allows belief in causality but this belief or intuition is independent of pure reason and it is this that led to the disparity between sensibility and intelligibility in early modern philosophy leading to the *Critique of Pure Reason*. Ginsborg argues that the issue at hand for all other early modern theories of abstraction was that they presupposed or depended upon "some antecedent recognition of a resemblance among the relevant ideas."[100] However, this is not the case with the account given by Aquinas.[101]

---

98 Ibid., 15.

99 Ibid., 16.

100 Ibid., 14.

101 It is possible that Hume was aware of Aquinas's account of empirical generality and drew upon Aquinas's account for his own; however, Hume failed to fully understand Aquinas's development of abstraction or Hume simply ignored the extent of the development. Following Samuel Taylor Coleridge, John K. Ryan argues that Hume borrowed without acknowledgment the "laws of association" from Aquinas' commentaries on the Parva Naturalia of Aristotle (i.e., the fifth chapter of Aquinas's commentary upon Aristotle's De Memoria et Reminiscentia). Coleridge discovered that Hume had various volumes of Aquinas with marginal marks and notes of reference in Hume's own handwriting and among these was the *Parva Naturalia*. Ryan explains that in the passages given by Aquinas in his commentary on Aristotle's *De Memoria et Reminiscentia* "St. Thomas has given a clear statement of the principles of association of ideas as evidenced in the process of recollection." See John K. Ryan, "Aquinas and Hume on the Laws of Association," *The New Scholasticism*, vol. XII, no. 4 (October, 1988), 366ff.

Gardeil explains, "the corner stone in the Aristotelian doctrine of intellectual knowledge is the theory of abstraction, for in this theory lies the answer to the crucial problem regarding the origin of such knowledge and its relationship to sense knowledge."[102] It was the *via antiqua* theory of abstraction following Aristotle that provided the philosophical glue between sensibility and intelligibility. It should be remembered that Aristotle introduced the theory of abstraction to counter the irreconcilable incompatibility between sensibility and intelligibility introduced by Plato, the Pythagorean, and the pre-Socratics.

Kant found it necessary to reject Hume's development of custom independent of reason and the developments of the rationalists and empiricists who had embraced the *via moderna* of Ockham, and thus Kant was left with the irreconcilable nature and distinction between sensibility and intelligibility to which Kant was to address the *Critique*. Thus Kant was forced, after having rejected the nominal theory of *abstraction* of the *via moderna*, to develop a nominal theory of transcendental idealism to explain how such unity between sensibility and intelligibility might be possible to save science, mathematics, and metaphysics from the irreconcilable differences between empiricists and rationalists.[103] One can

---

102 H. D. Gardeil, O.P., trans. by John A. Otto, Ph.D., *Introduction to the Philosophy of St. Thomas Aquinas III. Psychology* (London: B. Herder Book Co., 1956), 253.

103 Kant would thus later conclude in the *Critique*: "I see e.g. a spruce, a willow and a linden. In first comparing these objects among themselves, I notice that they are different from one another with respect to the trunk, the branches, the leaves and so forth; but now I go on to reflect only on what they have in common, the trunk, the branches, the leaves themselves; and I abstract from their size, shape and so forth; thus I receive [*bekommen*] a concept of tree" (*Logic*, §6, note 1; 9:24-25). Thus for Kant "to say that a concept is universal or general is to say that it is 'common to several objects' (Logic, §1, 9:91), and hence contrasts with an intuition, which is a singular representation.... There is, however, another, apparently distinct sense of 'universal' which is also invoked by Kant in describing the exercise of judgment... 'Universality' in this sense means, as he puts it, 'validity for everyone' (§8, 5:215)" (Ginsborg, 2003, 2). In the *Critique*, Kant would conclude that experience is the representations or sensible intuitions that come to us because of the way our senses are affected. Hence experience can only acquaint us with individual things since sensible intuitions are of the singular and never of the universal. Experience cannot acquaint us with properties or features held in common with other things, but in the "logical acts" of comparison, reflection, and abstraction one is able to abstract "from the features which differentiate them and [by] attending to the common features we arrive at the concept of tree, which presumably can be characterized as the concept of a thing with leaves, branches and a trunk" (Ginsborg, 2003, 6). See Hannah Ginsborg, *Thinking the Particular as Contained under the Universal*, (U. C. Berkeley: Dept. of Philosophy Draft, November 2003), 4-6, 11; also cf. Beatrice

see that this is the case as Kant continues in the same letter to address this very issue:

> In the *Dissertation* I was content to explain the nature of these intellectual representations ... as not being modifications of the soul produced by the object. But I silently passed over the further question, how such representations, which refer to an object and yet are not the result of an affection due to that object, can be possible. I had maintained that the sense representations represent things as they appear, the intellectual representations things as they are... if such intellectual representations are due to our own inner activity, whence comes the agreement which they are supposed to have with objects, which yet are not their products? <u>How comes it that the axioms of pure reason about these objects agree with the latter, when this agreement has not been in any way assisted by experience?</u>... when we ask how the understanding can form to itself completely *a priori* concepts of things ..., with which these things must of necessity agree, or formulate in regard to their possibility <u>principles which are independent of experience, but with which experience must exactly conform, -- we raise a question, that of the origin of the agreement of our faculty of understanding with the things in themselves,</u> over which obscurity still hangs.[104]

We see that as early as 1772 Kant raising the question of the *Critique*, how can experience exactly conform to *a priori* concepts that are independent of experience? How does it come about that *a priori* concepts about objects agree with the objects external to a knowing

---

Longuenesse, *Kant and the Capacity to Judge* (Princeton: Princeton University Press, 1998), 115-116; Robert Pippin, *Kant's Theory of Form* (New Haven: Yale University Press, 1982), 112ff; Hannah Ginsborg, "Lawfulness without a Law," 25 *Philosophical Topics* (1997): 53; Henry Allison, *Kant's Theory of Taste* (Cambridge: Cambridge University Press, 2002), 21ff. For a description of Okham's nominalism contrasted with various forms of medieval realism see Marilyn McCord Adams, *William Ockham* Vol.I, (Notre Dame: University of Notre Dame Press, 1987), 13-141.

104 Immanuel Kant, *Critique of Pure Reason*, trans. by N. Kemp Smith, 2nd ed. (London: Macmillan, 1929), 219-220.

subject, when this agreement has not been in any way assisted by experience?[105] The solution was not obvious given the early modern disparity between intelligibility and sensibility, and it is here that we turn to the difficulties connected with the concept of causation and Kant's debt to Hume. According to Robert Wolff, Kant was not acquainted with the *Enquiries* before the letter to Herz in 1772, but shortly after February 1772 there was a translation of James Beattie's *Essay on the Nature and Immutability of the Truth* (1770), which likely awakened Kant from his dogmatic slumber.

Beattie's treaties attempted to address Locke, Berkeley, and Hume as well as Descartes and Malebranche and thus Beattie quoted significant passages from the works of each author.[106] Beattie included the arguments from the *Treatise* and the *Enquiry Concerning Human Understanding* and thus Kant was made aware of a number of arguments including *the criticism of the causal maxim*.[107] Hume's skeptical attack on causality is contained in Part III of Book I of the *Treatise of Human Nature*, "Of Knowledge and Probability" and Section IV of the *Enquiry Concerning Human Understanding*, "Skeptical Doubts Concerning the Operations of the Understanding" given earlier.[108] Wolff indicates that

---

105 The issue that concerns Kant and the solution provided by the *via antiqua* is explained by Gardeil: "The proper object of the human intellect is the quiddity or nature of sensible things, which nature exists only in the singular, that is, in corporeal matter. The nature of a stone, for example, is such as to exist in this particular stone. Consequently, the nature of a stone or any material thing whatever cannot be known 'completely' and 'truly' unless it be understood as existing in the individual. The individual, however, is apprehended only by the senses or in the phantasms; and so for the intellect to come to know its proper object, it must of necessity turn to the phantasm, there to behold the universal nature existing in the individual...Thomistic philosophy, while uncompromising in affirming the essential and irreducible distinction between intellectual and sensory knowledge, is not less emphatic in its opposition to isolating and insulating these two forms of knowledge from each other...Though initially and essentially the faculty of the abstract and the universal, the intellect thus emerges as the faculty of the concrete individual as well" (H. D. Gardeil, O.P., trans. by John A. Otto, Ph.D., *Introduction to the Philosophy of St. Thomas Aquinas III. Psychology* (London: B. Herder Book Co., 1956), 155-156).
106 Robert Paul Wolff, *Kant's Theory of Mental Activity* (Gloucester, Mass: Peter Smith, 1973), 24.
107 Ibid., 24-25.
108 See footnote 71-74 esp. ft. 74 or Hume's argument against causality and necessity in the *Enquiry*. Wolff also mentions that it is likely that Kant was never aware of the *Treatise of Human Nature* prior to the writing of the *Critique* outside the work of Beattie's Essay. Kant also had a partial translation of Hume's other major works, which Hamann presented to Kant in 1780. Kant made use of some of the arguments from the *Dialogues Concerning Natural Religion* in the *Dialectic*, and later incorporated into the *Critique of Teleological Judgment*. Cf. Robert Paul Wolff, "Kant's Debt to Hume via Beattie," *Journal of the History of Ideas*, 21 (January-March 1960) : 117-123.

the *Enquiry* does not contain the causal maxim, which states *whatever begins to exist proceeds from some cause*. Hume gives the argument against the causal maxim in the *Treatise*.[109]

In the *Treatise* Hume accepted resembling, identity, relations of time and place, quantity, quality, contrariety, and causation as fundamental relations. However, Hume maintained that resemblance, contrariety, quality, and quantity were independent of time or place or position as understood in perception and therefore they could in fact provide a basis for science.[110] However, identity, time, place, and causation are not independent of the temporal order nor the relative positions in which they appear in perception according to Hume. Each of these do not depend upon the idea and as such they may be present or absent even if the idea remains the same. So, one can imagine two objects in a variety relations while the characteristics of the objects remain the same. On the other hand, we can judge a certain impression of two objects to be identical upon the conditions under which they appear.

Hume continues to argue that causality is itself dependent upon the temporal order or succession of cause and effect as well as spatial proximity. Thus a given relation can be destroyed by varying the appearance without changing the given ideas about the causal relation. One can judge a given relationship to be causally connected, but if spatially remote or separated in time, one can judge the same motions not to be causally connected.[111] Therefore causality is inseparable from spatial temporal relations. As will be seen shortly, this discovery became fundamental to Kant's philosophical program. Hume also held however, that since *ideas* nor identity nor time nor place can lead one beyond experience to knowledge of real existence and relations of objects, empirical *knowledge* is dependent upon cause and effect relations. Hume also found that a necessary *connexion* is an essential element in causal relations, but when one examines such relations one only finds

---

109 Robert Paul Wolff, *Kant's Theory of Mental Activity* (Gloucester, Mass: Peter Smith, 1973), 25.
110 Ibid., 26.
111 Ibid.

contiguity and succession between a cause and its effect. Again this analysis alters Kant's position laid out in the 1770 *Dissertation* leading to the development of the transcendental.

In the *Treatise of Human Nature*, the first question asked by Hume is how can it be said that every thing that has a beginning necessarily must have a cause, or simply *why is a cause always necessary*, or why is the causal maxim *whatever begins to exist, proceeds from some cause* is necessarily the case. The second, why should one conclude a particular cause must necessarily have a particular effect, and what inference should be drawn from a particular cause to its effect?[112] Hume argues that impressions of causal relations can be distinguished separately and it is possible to conceive that a necessary connection between a cause and its effect does not in fact exist. Further, since the relation of causal ideas are not themselves unalterable as might be the case with quantity for example, there is no reason to hold that these relations are intuitively certain. Using these arguments Hume maintains that one can imagine fire without heat, or the bodies striking one another without motion; Hume *argues at least it is possible that they so occur*.

It is here that Kant should have dealt with Hume's arguments directly, arguing that for such effects to exist a given cause is necessary, therefore it is necessary for what begins to exist, to have some cause. For fire to heat a given object, heat must necessarily exist in fire for without heat, a heated object heated by fire, cannot be heated. Therefore, both by definition and as demonstrated from experience, it necessarily follows that fire has heat because an object heated by fire must be heated to be heated and if heated by fire, fire must necessarily contain heat if the heated object is in fact heated by fire and it is impossible to imagine otherwise without error in judgment violating the law of contradiction.[113] For two bodies to strike one another, motion is

---

112 Ibid., 27.

113 For Aquinas, truth always implies a relation between the first mode of being and the second, but the relation can be considered either from the first mode of being which are things external to the knowing subject, or from the second mode of being, which is the knowing subject. If taken from the perspective of the first mode of being, truth is in things or a thing is true in proportion to its conformity with the intellect. One arrives at ontological truth in virtue of the first mode of being, the conformity of

necessary for without motion there is no movement from a potential accidental location to an actual accidental location and therefore for one object to be moved it must necessarily contain within its notion or accidental form the actuality of having been moved from one location to that of another.

It is by this accidental form as that by which the causal relation between two objects obtains, and as that by which the transfer of motion is apprehended by the intellect and affirmed in judgment. Therefore, a deductive demonstration based upon inductive necessity can be given for a necessary connection between a cause and its given effect. Thus the causal maxim can be apprehended and judged to exist by the intellect in the *via antiqua* of Aquinas. However, as indicated by Wolff, Hume argued in the *Enquiry Concerning Human Understanding* Part I that "particular causal inferences are not *a priori* or deductive in character, and hence lack the necessity which the rationalists ascribe to them."[114] And Part II argues against the uniformity of nature, "even after we have experience of the operations of cause and effect, our conclusions from that experience are not founded on reasoning, or any process of the understanding."[115] Hume attempts to demonstrate not only that the notion of causality cannot be *a priori*, but that the notion of a uniformity of nature likewise cannot be *a priori*.

Second Hume argues that there can be no uniformity between experiencing the operation of cause and effect and the process of understanding.[116] Because of the epistemological and metaphysical divide between the rationalists and empiricists, Hume indirectly engages the medieval understanding of *abstractio* or the uniformity of nature between experience and what can be known from experience using causality. Causality thus became the focal point for engaging the principle of the uniformity of nature between intellect and experience

---

thing to intellect: *adaequatio rei ad intellectum*. In the case of the second mode, truth is knowledge that bears a relation of conformity with its object, with some external reality. This second mode is logical truth, the conformity of intellect to thing: *adaequatio intellectus ad rem*.

114 Robert Paul Wolff, *Kant's Theory of Mental Activity*, (Gloucester, Mass: Peter Smith, 1973), 29.

115 As quoted by Robert Paul Wolff from the *Enquiry*, ed. By Selby-Biggs, p. 32

116 Robert Paul Wolff, *Kant's Theory of Mental Activity* (Gloucester, Mass: Peter Smith, 1973), 29.

in modern philosophy, but likewise in Medieval philosophy causality was central to what can be known from experience by the human intellect.

Hume had argued that all reasoning concerning matters of fact are *founded on the relation of Cause and Effect*, but no examination of a particular object can allow one to deduce the nature of its causal effect. One cannot determine from water, Hume argued, that water might suffocate. As Wolff explains, "Consequently, our knowledge of matters of fact is never based upon *a priori* reasoning. This, it is frequently said, is the challenge which Hume issued and which Kant answered. But Hume is here denying the *a priori* character only of individual causal judgments."[117] Wolff explains that Kant, however, *only demonstrated that whatever begins to be presupposes something on which it follows according to a rule* (*Critique*, A189) and therefore Kant never argued for particular causal relations such as heat from fire. [118] However, Kant's transcendental method does seem to apply not only to the general causal maxim, but also to particular causal events since any singular causal event is likewise conditioned by the *a priori* conditions of understanding.

As Wolff notes at the time of the letter to Herz in 1772 Kant began to realize the serious problems not addressed by the *Dissertation*, and the primary point of contention became the use of the pure concepts of reason which derive nothing from experience and yet are to lay claim to universal validity independent of knowledge of the noumena. Hume's attack on causal inference had undermined any attempt by Kant to limit pure concepts to phenomena while giving up any claim to the noumena. Wolff explains, "Hume's attack on causal inference undermined even this drastic retrenchment, for physics as well as metaphysics employed the concept of causation. Hence Kant was made to realize the need for a general critique of the functions of understanding and *a priori* knowledge."[119]

---

117 Ibid., 30.
118 Ibid.
119 Ibid., 31.

It was the argument in the *Enquiry* concerned with the particular causal inferences rather than the general maxim of causality that implied the problem of how synthetic judgments that are from experience can possibly be *a priori*. This led Kant to attempt to show how causal inferences could both be *a priori* and related to sensible experience. Wolff explains that the "Dissertation had separated sensibility from intelligence, restricting the first to phenomena and extending the second to things in themselves... the relation between intuitions and pure concepts had been treated only in a negative manner and with regard to the Leibnizian account of confused ideas. This however left Kant with a theory of two independent and coequal 'conditions of the possibility of experience,' sensibility and understanding."[120] Wolff continues by explaining that Hume's attack on causality forced Kant to examine more closely the relation between *sensibility and understanding*.

Wolff explains, "the key to Hume's attack is the dictum of the separability of distinguishables. But the only true invariable relations between causes and effects are spatial contiguity and temporal succession. Since objects in different parts of space or time are distinguishable, it follows that they are separable in the imagination, and consequently that the one can at least possibly occur without the other. Hume's analysis revealed that it was not causation *in addition to* space and time, but rather causation *as based on* space and time, which required a defense."[121] Thus was born the problem of the synthetic *a priori* and the need for the *Critique of Pure Reason*.

The *Critique* was an attempt to work out the relationship between the pure concepts of understanding such as space and time and *a priori* forms of intuition, and the relationship was set down in the *Deduction* and the *Analogies*.[122] Kant attempted to solve the incongruent relationship between the synthetic and the *a priori* by developing the theory that pure concepts are actually rules for the *synthesis of a manifold*

---

120 Ibid.
121 Ibid.
122 Ibid., 32.

of intuition. The following section will discuss various problems with Kant's solution to the synthetic *a priori* distinction introduced in opposition to the Humean critique on causal inference. Kant's transcendental philosophy is an attempt to resolve the dissociation between sensibility and intelligibility as introduced by Cartesian dualism, preconditioned by the nominalism of the *via moderna,* and further developed by the irreconcilable differences between eighteenth century continental rationalism and British empiricism that had taken the *via moderna* to its logical conclusion resulting in a complete disconnect between sensibility and intelligibility.

Chapter 3

THE PROBLEM OF THE SYNTHETIC A PRIORI

Kant understood that Hume had argued that the synthetic *a posteriori* relation is only empirical and thus the representation is therefore never *a priori* therefore the *a posteriori* can never lead to necessity or universality. Kant was awoken from his dogmatic slumber upon realizing the problematic introduced by the empiricist and rationalist debate: a synthetic *a posteriori* determination of causality itself as the basis for science can never be grounded in the *a priori*. Therefore neither science nor metaphysics had a philosophical basis in universality and necessity for doing either science or metaphysics without a critique of the bounds of reason. Kant argued that metaphysics is a *specie* of synthetic judgments and by examining the synthetic he would be able to shed light on the scope and limits of metaphysical judgments.

Kant explains: "the proper problem of pure reason is contained in the question: How are *a priori* synthetic judgments possible? That metaphysics has remained in uncertainty and contradiction is due to this problem and because no one has considered the distinction between analytic and synthetic judgments. Upon the *solution to this problem, or upon a sufficient proof that the possibility which it desires to have explained does in fact not exist at all*, depends the success or failure of metaphysics" (emphasis added).[123]

---

123 Immanuel Kant, *Critique of Pure Reason*, trans. by N. Kemp Smith, (London: Macmillan, 1929), 55.

Although Kant had seen that the solution which synthetic *a priori* judgments or a critique of pure reason had hoped to explain might not exist at all, yet for Kant the success or failure of metaphysics came to depend entirely upon a solution to the problem of the synthetic *a priori*, answering the question how are synthetic *a priori* judgments possible and thus what are the limits of pure reason? Kant had concluded that Hume had seen one aspect of a much larger epistemological problem. The rationalist philosophy of Leibniz and Wolff on the one hand and the British empiricism of Newton, Samuel Clarke, and Hume on the other introduced a difference between the "sensible and the intelligible as merely logical, giving a completely wrong direction to all investigations into the nature and origin of human knowledge."[124]

Kant concluded that this difference is in fact transcendental and does not merely concern logical form, it concerns the very origin and content of human knowledge. [125] Kant thus introduced the transcendental into

---

124 Ibid., 84.

125 It is here that one finds the full implication of the Kantian turn in epistemology and the problem that arose from the empiricist and rationalist debate. The rationalist and empiricist debate based upon nominalism introduced a dichotomy between the sensible and intelligible in logic as well as in the nature and content of human knowledge. The seeds of the modern break in noetics may have been expressed in a different form by earlier forms of formal logic such as Apuleius's *Peri Hermeneias*. According to Sullivan, "Apuleius's *Peri Hermeneias* and its derivative abstracts in Martianus, Cassiodorus, and Isidore, Boethius's *De Syllogismis Categoricis*, and Boethius's translation of Aristotle's *Prior Analytics*, provide the only known ancient Latin sources of (more or less detailed) categorical syllogistic theory which might have influenced medieival thinkers up until the time of Peter Abelard" (Mark W. Sullivan, *Apuleian Logic The Nature, Sources, and Influence of Apuleius's Peri Hermeneias* (Amsterdam: North-Holland Publishing Company, 1967), 190). Sullivan continues by adding that "Boethius's *De Syllogismis Categoricis* exerted little influence on the logic of the Latin West from the sixth to almost the end of the eleventh century, but the *Peri Hermeneias* of Apuleius was one of the few texts known on syllogism being used not only for gaining knowledge of syllogism but also in order to better understand the *Peri Hermeneias* of Aristotle" (Ibid., 191). Sullivan indicates that it is only by the twelfth century that Apuleian syllogistic theory was replaced by the New Logic of Aristotle by the success of the *Sullogismis Categoricis* as the *principal* ancient source in the *Analytics* of Aristotle (Ibid., 192). However, "by the end of the twelfth century, Apuleius's *Peri Hermeneias* continues to be studied in the schools along with Boethius's *De Syllogismis Categoricis* and the New Logic of Aristotle" (Ibid., 192-208). As precursor to contemporary *a priori* formal logic it may have influenced the Middle ages through Boethius since Boethius likewise drew from Apuleius's *Peri Hermeneias*. Sullivan provides a detailed comparison between Apuleius's *Peri Hermeneias* and Boethius's *De Syllogismis Categoricis* and concludes "Boethius's De Syllogismis Categoricis was influenced directly or indirectly by Apuleius's Peri Hermeneias... it becomes quite probable that Boethius had a copy of Apuleius's treatise at hand when he composed De Syllogismis Categoricis.... we regard it as probable that Boethius borrowed directly from Apuleius's *Peri Hermeneias*" (Ibid., 225, 227; see 207-225). Sullivan also attempts to make the case that Peter Aberlard's *Dialectica* was likewise influenced indirectly by Apuleius's *Peri Hermeneias* through Boethius's *De Syllogismis Categoricis* (Ibid., 228).

One distinction between Aristotelian logic and that of Apuleius is how the proposition is to be understood. Sullivan explains that for Apuleius (c. 125) "the subject matter of the art of disputing (*ars disserendi*), of logic, is the proposition (propositio) and that a proposition is a declarative utterance (oratio pronuntiabilis), a special kind of speech. Aristotle, according to Bochenski, holds that the subject of logic is the syllogism and that a syllogism is logos (of a special sort)...." (Mark W. Sullivan, *Apuleian Logic The Nature, Sources, and Influence of Apuleius's Peri Hermeneias* (Amsterdam: North-Holland Publishing Company, 1967), 146). Sullivan indicates that according to Bochenski, the demonstration for Aristotle is not about words but about things of the soul, and laws hold about 'spoken affirmations' because similar laws hold with regard to the 'judgments of the mind' and hence "for Aristotle logic is primarily an affair of right thinking and secondarily, a matter of correct speaking. Not so for Apuleius... logic is primarily a matter of right speaking and secondarily (it seems) an affair of correct thinking... the logician, for Apuleius, is concerned primarily with the relations between propositions, between declarative utterances, and not with those holding between the complete senses or thoughts which propositions may express. Logic is a linguistic, not a psychological, discipline for Apuleius; his position on this matter appears to be unambiguous" (Ibid, 146-147). For a discussion of Sullivan's account and that of Grabmann see (Mark W. Sullivan, *Apuleian Logic The Nature, Sources, and Influence of Apuleius's Peri Hermeneias* (Amsterdam: North-Holland Publishing Company, 1967), 228-232. In opposition to Sullivan's account of Boethius see Taki Suto, "Boethius on Mind and Language: For a Study of Boethius' Commentary on Peri hermeneias" http://www.hmn.bun.kyoto-u.ac.jp/report/2_tetsugakui/2_16.pdf. Also see John C. Magee, *Boethius on Signification and Mind*, (Leiden: Brill, 1989); Mark Sullivan, "What was True or False in the 'Old' Logic", *Journal of Philosophy*, 22, 67 (Oct. 1970): 788-799.

Logicians by the end of the twelfth century in adopting the New Logic of Aristotle began to move away from Apuleius's *Peri Hermeneias* and returning to an Aristotelian moderate realist epistemology in logic. One finds such a shift in Peter Abelard (d. 1142) as well as in Peter Helias who was the founder of the medieval logic of language and professor at Paris around 1140. Peter Helias in explaining the Priscian's grammar text *Institutiones grammaticae* using Aristotle's *Categories* and *On Interpretation* taught that parts of speech are distinguished by their *modi significandi*, which was Peter Helias's understanding of Priscian's *proprietates significationum* (Greggory, Rocca, O.P., "Res Significata and Modus Significandi", *Thomist* , 55(1991), 177). The modus is a way of signifying or consignifying something where the proposition indicates the mind's concept (Ibid.). Hence a noun signifies a "substance with quality" for Peter Helias and thus a noun was not merely a name but the name signified something in existence (Ibid.). Between 1270-1350 the *grammatica speculativa* was developed by those called the *Modistae*. Rocca explains, "Relying on Aristotle's logical corpus (as interpreted by Boethius), Priscian, Donatus, and Peter Helias, the *Modistae* treatises merged grammar with logic and even metaphysics, attempting to construct a philosophy of language that could describe a universal linguistics and grammar that are isomorphic with and dependent upon common reality. The Modistae take for granted a realistic epistemology" (Ibid, 178).

The *Modistae* introduced the *Tractatus de modis significandi* or *Summae modorum significandi* (Ibid., 178). The various work by the *Modistae* provided a realist epistemological account using the triad of *modus essendi*, *modus intelligendi*, and *modus significandi*, such that the *modus esendi* grounds *res*, represented by the *conceptus*, which is grounded in the *modus intelligendi*, and signified by the *dictio*, which is grounded in the *modus significandi* (Ibid., 178). Rather than the categorical syllogism or proposition simply being the *dictio* as maintained by Apuleius's *Peri Hermeneius*, the proposition was the simple or complex concept, which was grounded in *res* (objective things) and was signified by the *dictio* (i.e., declarative utterances). By having a particular *modus significandi*, a given vocal sound becomes a word and a part of speech in a given language, but the *modus significandi* is conditioned by the way in which intellectual knowledge grasps reality intentionally (*modus intelligendi*), and intellectual knowledge itself is conditioned by the various categories and kinds of reality or existence (*modus essendi*). Aquinas recognized the traditional triad (*Herm.* 1.8.90) and the realist epistemology, and Aquinas employed the *modus significandi* which became identified with a word's *consignificatio* allowing the intelligible depth of a concept to signify the individual notes of a thing in comprehension or connotation (cf. Greggory, Rocca, O.P., "Res Significata and Modus Significandi", *Thomist* , 55(1991), 177-180.)

the empiricist and rationalist debate having inferred: "it is not that by our sensibility we cannot know the nature of things in themselves in any save a confused fashion; we do not apprehend them in any fashion whatsoever. If our subjective constitution be removed, the represented object, with the qualities which sensible intuition bestows upon it, is nowhere to be found, and cannot possibly be found. For it is this subjective constitution which determines its form as appearance."[126]

Modern philosophy had come to accept that by our sensibility one cannot know the nature of things in themselves except in a confused fashion. It therefore followed that if the subjective constitution whereby a given object was removed, the object itself could not be found. Kant realized that the subjective constitution is that which determines its form as appearance, but there was no connection between the subjective constitution and experience and it is this disconnect that the transcendental schemata and the analogies in the *Critique of Pure Reason* hoped to bridge.

Kant seen the possibility that David Hume was the nearest to seeing the problem, but Hume had focused on synthetic propositions, and Hume believed that he had shown that *a priori* proposition are impossible.[127] If this is accepted, Kant argued, "all that we call

The *Modistae* triad diminished in the wake of nominalism since nominalism found that the *dictio* grounded in the *modus significandi* was sufficient to explain declarative utterances which comprehend meaning Nominalism could not accept the realism of universals, the assimilation of intelligible forms abstracted from *res*. Hence the *via moderna* emphasized the role of *dictio* and the *modus significandi* while ignoring the roles of the *modus essendi* and the *modus intelligendi*. Consequently propositions were understood to be declarative utterances (*orationes pronuntiables*) that comprehend meaning rather than propositions being understood as concepts that signify quiddities or the essence of things and where words signify propositions so understood. It is this nominalism, where propositions were considered declarative utterances that comprehend meaning rather than concepts that signify the essence of things, that Kant inherited through Wolffian logic and that provided the basis for Kant's transcendental analytic and synthetic *a priori* judgments. (for a detailed discussion of the *Modistae* and developments in logic see Martin Grabmann, *Mittelalterliches Geistesleben*, 3 vols. (Munich: M. Hueber, 1926; 1936; 1959), 1:115-46; also vol. 1, chap. 4, and vol. 3, chaps. 3 and 12.

126 Immanuel Kant, *Critique of Pure Reason*, trans. by N. Kemp Smith, (London: Macmillan, 1929), 84.

127 Kant believed that Hume had demonstrated that synthetic propositions could not be *a priori* and if this were the case, there is no necessary causality at the level of experience, ontology, or synthetic judgments. If this were the case, Kant reasoned, then there is no basis for metaphysics or science or even mathematics. The problematic itself arose from the empiricist and rationalist debates that presupposed nominalism and therefore were unable to make a proper connection between experience and reason: a disconnect between reason and experience had occurred. Therefore if one rejects Kant's transcendental

metaphysics is a delusion."[128] For Kant the problem of metaphysics, i.e., pure reason is at the level of determining how synthetic *a priori* judgments are possible, and analytic judgments as the analysis of concepts that inhere in reason *a priori* is only "a preparation for metaphysics proper, that is, the extension of it's *a priori* synthetic

---

solution, it remains necessary to demonstrate how *a priori* synthetic judgments are in fact possible, i.e., that the rationalist and empiricist problematic is not actually a problem exactly because of the failure of nominalism to account for *modus essendi* and *modus intelligendi*. In this account it is necessary to demonstrate that synthetic judgments of causality, substance, and the other categories can be both universal and necessary and yet be apprehended from experience such that no disconnect between reason and experience exists.

Since Kant maintained that Hume believed that he had demonstrated that synthetic propositions cannot be *a priori*, the following emphasis is upon the synthetic *a priori* rather than *a posteriori* propositions or judgments since we are addressing Kant's attempt at resolving Hume's critique. For Kant the *a posteriori* was not the issue since Kant was concerned with universality and necessity of synthetic judgments and Kant readily admitted with Hume that the *a posteriori* could not be universal or necessary. If one were to argue for universality and necessity using the *a posteriori*, one would necessarily be forced to address Hume's outdated psychology of how one obtains custom and belief through the *a posteriori* since it seems that custom is the sole means by which one attains belief in necessity and causality according to Hume and one would be forced to develop an *a priori* understanding of the *a posteriori* but independent of the Kantian transcendental. Independent of the transcendental in the sense of being independent of the rules for the imaginative synthesis of the manifold or simply independent of the rules or schemata some of which are empirical and others are *a priori*. Since Aquinas had no notion of Hume's *a posteriori* development of a disposition of custom and belief as the sole means by which one infers necessity and efficient causality entirely independent of the *a priori*, and because Aquinas did not have the Kantian transcendental notion of the *a priori*, the following proposal attempts to resolve both issues and the fundamental problem introduced by the rationalist/empiricist debate by demonstrating that synthetic *a priori* judgments can be apprehended from experience and yet remain both universal and necessary (i.e., *a priori* in nature) thus answering how synthetic judgments can be *a priori* while demonstrating that the synthetic *a priori* problem does not in fact exist. It is through this *a priori* synthetic reformulation that one can obtain knowledge of efficient causality from experience that is both universal and necessary.

This solution, however, does not force *a priori* synthetic judgments to be *a posteriori* since the *a posteriori* is limited to the domain of sensibility or experience independent of reason. Based upon Hume's account of the *a posteriori* the *a posteriori* is essentially Aquinas's notion of sensibility and Aquinas agrees with Hume that the intellect is, in a qualified sense, independent of sensibility or experience and thus one must return to the phantasms in apprehending singulars in Aquinas, but contrary to Hume the intellect can still apprehend universality and necessity from experience or sensibility even though a distinction exists between experience and the intellect. For Kant synthetic judgments are related to experience but Kant's rationalist approach to Hume's critique of reason introduced the *a priori* transcendental philosophy to answer *how* it is possible for the intellect to apprehend universality and necessity in relation to experience or sensibility in synthetic judgments. Since Aquinas was able to abstract universality and necessity from experience or sensibility independent of the Kantian transcendental, it is necessary to address the Kantian synthetic *a priori* and the transcendental turn developed by Kant. Independent of the transcendental in the sense of being independent of the rules for the imaginative synthesis of the manifold or simply independent of the rules or schemata some of which are empirical and others are *a priori*.

128 Immanuel Kant, *Critique of Pure Reason*, trans. by N. Kemp Smith, (London: Macmillan, 1929), 55.

knowledge."[129] In discussing the general problem of pure reason, Kant suggests that an analysis of concepts is useless since it only tells of the content of these concepts but not how the concepts are arrived *a priori*.[130] Hume had shown that synthetic propositions cannot be *a priori* and if this is the case, there can be no necessary causality at the level of experience or synthetic judgments.

Kant explains: "Hume occupied himself exclusively with the synthetic proposition regarding the connection of an effect with its cause (*principium causalitatis*), and he believed himself to have shown that such an *a priori* proposition (*whatever begins to exist, must have a cause of existence*) is entirely impossible.[131] If we accept his conclusion, then all that we call metaphysics is a mere delusion whereby we fancy ourselves to have rational insight into what, in actual fact, is borrowed solely from experience, and under the influence of custom has taken the illusory semblance of necessity."[132] Kant argued that if this is the case, there is no basis for metaphysics since metaphysics is grounded

---

129 Ibid., 57.

130 Ibid.

131 Hume explains that the necessity of a cause is assumed to be the case and typically not demonstrated: "To begin with the first question concerning the necessity of a cause: It is a general maxim in philosophy, that whatever begins to exist, must have a cause of existence. This is commonly taken for granted in all reasonings, without any proof given or demanded. It is supposed to be founded on intuition, and to be one of those maxims which, though they may be denied with the lips, it is impossible for men in their hearts really to doubt of. But if we examine this maxim by the idea of knowledge above explained, we shall discover in it no mark of any such intuitive certainty; but on the contrary shall find, that it is of a nature quite foreign to that species of conviction." (David Hume, *A Treaties of Human Nature,,* 1911 Everyman's Library Edition. Bk. 1 Pt. 3 Sec. 3 Para. 1/9, p. 78). However, in a letter to John Stewart on Feb. 1754 in reaction to an essay written by Stewart where Stewart indicates that Hume's skeptical philosophy rejects the notion "that something may begin to exist, or start into being without a cause," Hume explains that he does not deny that things arise from a cause nor does he deny their certainty, Hume only denies that our certainty of such propositions can proceed from intuition or demonstration. Hume writes: "I never asserted so absurd a Proposition as that any thing might arise without a Cause: I only maintain'd, that our Certainty of the Falshood of that Proposition proceeded neither from Intuition nor Demonstration; but from another Source. That Caesar existed, that there is such an Island as Sicily; for these Propositions, I affirm, we have no demonstrative nor intuitive Proof. Would you infer that I deny their Truth, or even their Certainty? There are many different kinds of Certainty; and some of them as satisfactory to the Mind, tho perhaps not so regular, as the demonstrative kind" (Hume: TLDH Vol 1 Ltr 91 To: Stewart [54] p 186). In opposition to Hume, the rationalist position of Kant would attempt to demonstrate that causal necessity is *a priori* and this by demonstrating how one can have universal causal necessity independent of experience but consistent with possible experience.

132 Immanuel Kant, *Critique of Pure Reason*, trans. by N. Kemp Smith, (London: Macmillan, 1929), 55.

upon the need for necessary causality. However, although Hume held that demonstration was not possible, Kant held that it was necessary to demonstrate how *a priori* synthetic judgments are in fact possible thus introducing the critique of pure reason.[133]

Kant drew a clear distinction between those disciplines that can yield *a priori* knowledge, such as mathematics and physics, and those that Kant believed could not be demonstrated by "traditional metaphysics."[134] For Kant, God, the immortality of the soul, etc. *were needed as things to act morally* and are to be *assumed by faith*. Kant thus limited knowledge of God or any metaphysical demonstration of God's existence insofar as such knowledge was to be based upon speculative reason. In so doing Kant believed he had made room for faith while expressing the full scope of practical reason.

This distinction rests upon Kant's general claim that knowledge requires both concepts and intuitions, that coherent knowledge-claims can only be made about objects one can in principle perceive. Specifically we can only make coherent knowledge-claims about objects that are spatial and temporal, because space and time are the pure forms of human intuition. Thus synthetic *a priori* judgments express the essential features of knowledge of objects and the limits of such knowledge. To go beyond the limits by talking about objects that are non-spatial and non-temporal, objects that cannot possibly yield intuitions, is to talk nonsense. Therefore we find that mathematics and physics yield genuine knowledge because they are concerned with objects of possible experience. However, to talk about God who is neither spatial nor temporal, is to talk nonsense.

---

133 Likewise, the significance of Kant's understanding of Hume's position can be illustrated by Kant's assertion that if Hume "had recognized that, according to his own argument, pure mathematics, as certainly containing *a priori* synthetic propositions, would also not be possible" then from such an assertion, Hume's "good sense would have saved him" (*Critique*, 55). In the *Critique of Pure Reason*, Kant is primarily concerned with saving science, mathematics, and metaphysics from Hume's rejection of the possibility of synthetic *a priori* propositions.

134 T. E. Wilkerson, *Kant's Critique of Pure Reason, A Commentary for Students*, (Oxford: Clarendon Press, 1976), 101.

So the *Analytic* and *Aesthetic* are concerned with preserving the genuine sciences of mathematics and physics. However, the Dialectic is devoted to those who wish to go beyond the limits of synthetic *a priori* judgments in speculative metaphysics and Kant hopes to expose the mistakes of philosophers such as Descartes and Leibniz, who attempted to make claims about objects that are not and could not be objects of possible experience, thus objects such as the soul, the limits of the universe, and God are not coherent knowledge-claims since they are neither spatial nor temporal. As such these objects are illusions of speculative reason, but these objects apply at the level of practical reason. Kant "wants to expose 'the logic of illusion' (A293/B349), which consists in using concepts to which no intuitions correspond, concepts of objects that are not spatial and temporal."[135]

The Kantian Synthetic a priori Defined

Kant held that there are two possible subject predicate relationships in judgment. The first was the analytic judgment where the predicate is contained in the subject by definition such as "all black dogs are black" or the "whole is equal to its parts". The second is the synthetic judgment where the predicate is an addition to the subject, and where the predicate is not necessarily part of the subject. These include factual statements, knowledge from observation, hypothetical or modal judgments, categorical judgments, and *a posteriori* judgments. For example, 'these roses are red' or 'if the axe is made of iron then it is hard'.

For Kant there can be synthetic *a posteriori*, synthetic *a priori*, and analytic *a priori*, but not analytic *a posteriori* judgments as illustrated in Fig. 1. There can be no analytic *a posteriori* judgments because all analytic judgments are by definition *a priori*. By nature of being analytic, analytic judgments are *a priori*. Just as there is a separation between sensibility and intelligibility, Kant maintains that cognition of

---

135 Ibid., 101.

an object consists in the concept and intuition. All empirical intuitions from experience are synthetic *a posteriori* when the manifold of intuitions are united in judgment but are not universal and necessary. Therefore for Kant no synthetic *a posteriori* judgment can be *a priori* following Hume's critique of the *a priori*.

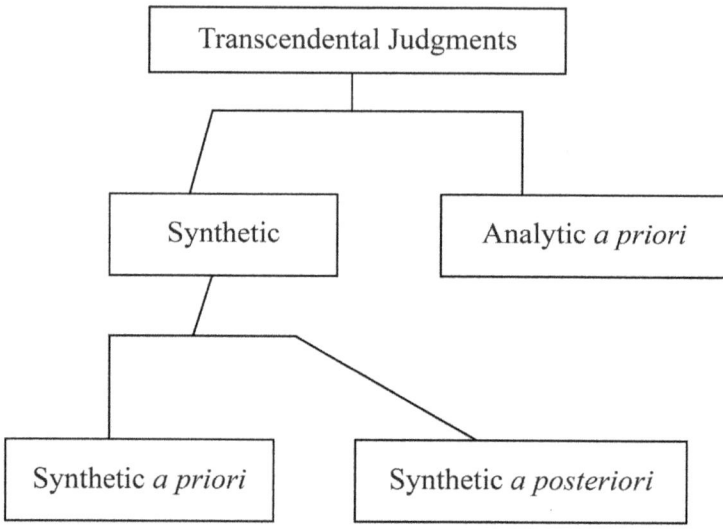

Figure 1 Judgments in Transcendental Logic

Kant realized that Hume's critique made it impossible for synthetic *a posteriori* judgments to be *a priori*.[136] Kant explains that knowledge that

---

136 This is of course why the noetic of Aquinas cannot be *a posteriori*. For Aquinas synthetic judgments abstracted from experience are in fact universal and necessary and are therefore *a priori* but by definition the *a posteriori* cannot be universal or necessary according to Hume and Kant. Although it is true that Kant maintained that the *a priori* is independent of experience, but this was conditioned by the Humean critique that maintained that reason cannot obtain in any way universality or necessity from experience. Kant following Hume's critique correctly seen that experience or sensation can never judge of necessary universal relationships, but both Hume and Kant failed to see that it is possible for the intellect to abstract universality and necessity from experience at one stage and then make judgments about simple concepts of universality and necessity at another. Hume maintained that custom was therefore required because the intellect could never infer from experience such relationships as causality, necessity, universality and the like because custom makes possible the belief that these relationships exist. However, Kant realized that this would defeat any notion of the *a priori*, and Kant proposed that synthetic judgments had to in some way be associated with possible experience. Kant proposed that such synthetic judgments must therefore be determined by and conditioned according to the *synthetic unity of apperception* or the *unity of consciousness* which is simply understanding itself understood in terms of

is independent of experience and independent of all impressions of the senses is entitled *a priori*, and this species of knowledge is distinct from empirical, which has its sources *a posteriori*, that is, in experience.[137] For Kant *a priori* knowledge is not simply knowledge independent of this or that experience, but knowledge absolutely independent of all experience and not derived from any experience.[138] Thus for Kant, the *modes of knowledge* are of two forms, knowledge independent of experience (*a priori*) and knowledge dependent upon experience (*a posteriori*). Pure *a priori* concepts are those which cannot be *derived* from experience, and necessity and strict universality are the "criteria of *a priori* knowledge, and are inseparable from one another."[139] A judgment is thought with strict universality where no exception is allowed and it is *a priori* when it is not derived from experience, but is valid absolutely *a priori*.[140] Substance, for example, is an *a priori* concept for Kant because once one removes all empirical concepts of an object, one cannot take away that property through which the object is thought as substance although substance as such is never perceived.

---

the transcendental philosophy for Kant. However, this approach failed to solve the original problem of the disparity between sensibility and intelligibility introduced by the Rationalists and Empiricists in a manner consistent with the natural mode of human knowing.

Kant imposed upon the manifold of intuition the forms of sensibility associated with the categories. The categories are connected with the forms of sensibility through the transcendental synthesis of the imagination. Therefore the *a priori* is imposed on the empirical content of sensibility rather than the empirical content of sensibility and the objects of possible experience being that from which the essence or quiddity as universal and necessary are abstracted. Hence one must first know the stone before experiencing the stone in Kant's account, which is absurd. Or, one must first have the schema of causality and necessity prior to apprehending causality and natural necessity from existence, which is likewise impossible. Again, this artificially dictates that causality must exist in the synthetic unity of apperception prior to it being apprehend by the senses and from sensible conditions external to the understanding. One must have a synthetic notion of causality conditioned by the schema and categories prior to being able to understand that billiard ball A was the cause of motion of billiard ball B upon being struck by billiard ball A, which is wrongheaded. However, contrary to Hume, if all is removed except billiard ball A and B, and B is struck by billiard ball A, the motion of B will be what determines the intellect to the cause of B's motion in one single causal event. Hume mistakenly rejects this possibility altogether, and Kant mistakenly accepts Hume's critique and thus altogether reverses the normal order of human knowledge acquisition contrary to both Aristotle and Aquinas.

137 Immanuel Kant, *Critique of Pure Reason*, trans. by N. Kemp Smith, (London: Macmillan, 1929), 42-43.

138 Ibid., 43.

139 Ibid., 44.

140 Ibid., 44.

Owing to the necessity with which the concept of substance forces upon us, substance has its "seat in our faculty of *a priori* knowledge."[141]

Some *a priori* judgments can be either analytic or synthetic, but there are no analytical judgments that are empirical. Kant defines *analytic judgments* as those in which the connection of the predicate with the subject is through identity; those in which this connection is thought without identity is entitled synthetic.[142] For example Kant considers, 'all bodies are extended,' as analytic because extended is bound to the subject, i.e., body. Likewise, identity is not that of equality but rather that of a predicate contained in its subject for example 'all black dogs are black' where the predicate 'black' is contained in the subject 'black dogs'. All 'black dogs' are necessarily and universally 'black' because the predicate is necessarily contained in the subject. One can simply analyze the concept by becoming aware of the manifold one thinks in that concept to see that the predicate is contained in the subject in the case of analytic judgments. *Synthetic judgments* on the other hand is where the predication of a proposition is not contained in the subject. In the proposition, 'all bodies are heavy', the predicate is different from anything one thinks in the concept body since by all bodies are not by definition heavy; and the addition of such a predicate or an extension therefore yields a synthetic judgment.

Kant then argues that judgments of experience are synthetic judgments because in framing the judgment one appeals to something in experience outside the concept in its support. Analytic judgments in contrast have no need to appeal to anything outside the concept in its support because the concept already contains all the conditions required for the judgment.[143] Analytic judgments occur in virtue of first principles of speculative reason such as the principle of contradiction: "I have only to extract from the concept, in accordance with the principle of contradiction, the required predicate, and in so doing can

---

141 Ibid., 45.
142 Ibid., 48.
143 Ibid., 49.

at the same time become conscious of the necessity of the judgment--
and that is what experience could never have taught me."[144]

In contrast to analytic judgments, in the case of synthetic judgments, one extends his or her knowledge of the concept by attaching to the concept of body the notion of extension, size, shape, etc., and by looking back at the experience from which these notions were derived. The association of the predicate with the subject is contingent upon an experience which is itself a synthetic combination of intuitions.[145]

In the case of *a priori* synthetic judgments, however, although synthetic judgments are supposed to be derived from experience, experience is not available to go beyond a given concept in the case of *a priori* concepts or *a priori* judgments.[146] For example, Kant proposes that the proposition 'everything which happens has its cause' is not based upon experience but is still a synthetic judgment. The predicate 'cause' lies outside the subject 'that which happens,' and therefore the predicate is not contained in the subject so it is a synthetic rather than an analytic judgment. The synthetic judgment is *a priori* in virtue of strict universality and necessity.[147] But the question arises, how can one associate the *a priori* that is independent of experience with a predicate that is supposed to be derived in some way from experience. How can a synthetic judgment possibly be *a priori* and not *a posteriori*?

The Kantian Synthetic a priori Solution

The problem addressed by the *Critique of Pure Reason* was how synthetic *a priori* judgments could be possible? The answer in short is that all empirical content of sensibility are synthetic *a priori* when the

---

144 Ibid., 49.

145 Ibid., 50.

146 Ibid., 50.

147 In synthetic *a priori* judgments experience does not give support to the understanding when it believes that it can discover outside the subject a predicate not contained in the subject, yet still considered to be connected to it. The support is not experience "because the suggested principle has connected the second representation with the first, not only with greater universality, but also with the character of necessity, and therefore completely *a priori* and on the basis of mere concepts" (Immanuel Kant, *Critique of Pure Reason*, trans. by N. Kemp Smith, (London: Macmillan, 1929), 51).

manifold of intuitions are united through the *synthesis of apprehension* under a concept united to the *forms of sensibility* through the categories in *the transcendental synthesis of the imagination* causing the *empirical content of sensibility* to be an *a priori* concept that is universal and necessary.

The essential issue is that *a priori* synthetic judgments provide the basis for Kant's Copernican revolution. For Kant the question was how can one have *strict universal* and *necessary* judgments where the *predicate is not contained in the subject without appealing to experience?* By justifying strict universal and necessary judgments that permit the predicate not to be contained in the subject while avoiding an appeal to experience, Kant believes that he can circumvent Hume's critique and save the scientific enterprise by providing a trustworthy basis for pure understanding. Kant argues that by determining how *a priori* synthetic judgments can occur, one provides a sure and trustworthy basis for knowledge yielded by pure understanding.[148] One aspect to resolving the question of *a priori* synthetic judgments was for Kant to make a distinction between an *empirical perspective* and a *transcendental perspective* permitting Kant to develop *a priori* conditions of the understanding independent of experience.

For Kant the *transcendental perspective* defines the form of appearance as that which so determines the *manifold of appearances* that it allows of being ordered in certain relations. The *form of appearance* are *a priori* rules that govern the exercise of the *cognitive power or the understanding*. Appearance of sensible impressions, although they are representations of objects, they are not knowledge of things-in-themselves. Appearances are objects of knowledge conditioned by the rules that govern the exercise of *vis cogitativa*. The *empirical perspective* holds that appearances that correspond to sensation are given to us *a posteriori*.

Kant explains that *a priori* synthetic judgments are possible because we have cognitive capacities or powers and these cognitive faculties

---

148 Immanuel Kant, *Critique of Pure Reason*, trans. by N. Kemp Smith, (London: Macmillan, 1929), 51.

have *a priori* conditions which are themselves *a priori* rules that govern the exercise of these cognitive powers. To think of an object of possible experience is just to connect one's perception according to a given rule. This connection between an object of possible experience and a given rule in turn makes the object of possible experience subject to an *a priori* rule, and this subjection of possible experience to an *a priori* rule is a transcendental condition of experience where experience is to be understood as knowledge by means of connected perceptions. Kant argues in the *First* and *Second Analogy* that possible experience is subject to *a priori* rules that determine the existence of objects of possible experience and that determine the transcendental condition by which synthetic knowledge can be considered *a priori*.

Therefore in general terms Kant's solution to the synthetic/*a priori* distinction is simply that objects of possible experience are subject to *a priori* rules that are the transcendental condition by which synthetic knowledge is *a priori*.[149] Henry Allison in commenting on the *First*

149 It should be noted that for Kant appearances attained *a posteriori* can never be universal or necessary, thus the universal abstracted from the particular in the *via antiqua* can never be *a posteriori* by the Kantian definition even if universality is taken in terms of extensionality in accord with the *via moderna*. Universality and necessity can only be attained in the analytic or synthetic *a priori* distinction and never *a posteriori*. Therefore, Thomists who choose to use *a priori* and *a posteriori* terminology must by definition use synthetic *a priori* in all cases to attain the universal and necessary conditions imposed upon experience explicit in the Kantian term *a priori* since what is known as universals by definition in the *via antiqua*. Thomists and analytic philosophers must either not use the terms *a posteriori* and *a priori*, or use such terms with an understanding of the implicit notion of Hume's critique and development of the *a posteriori* or the transcendental conditions imposed by the *a priori* by Kant thus one cannot attain the universal from the *a posteriori* of Hume nor abstract the universal from experience in the transcendental philosophy of Kant, or one must redefine the *a posteriori* explaining how in opposition to Hume one can have an *a priori* notion from the *a posteriori*, or one must redefine the *a priori* removing the Kantian transcendental. Redefining abstraction in terms of the *a posteriori* is an attempt to circumvent the need for synthetic judgments and the transcendental while failing to address the problem that synthetic *a priori* judgments were intended to resolve after the early modern shift. The use of the *a posteriori* by Thomists only exasperates the modern problem since nothing *a posteriori* can be universal or necessary from the perspective of modern Kantian and Humean philosophy, nor in Aquinas if the *a posteriori* is understood in Humean terms of being independent of reason.

Therefore the problem of synthetic *a priori* judgments must be addressed by Thomists who hold that one can attain universality and or necessity by abstracting the universal from sense knowledge of sensible things. The early modern turn was based upon a false problem addressed previously in the history of philosophy. A problem caused by a disparity between intelligibility and sensibility. The Kantian transcendental solution and the Humean notion of the *a posteriori* are both likewise inconsistent with the natural mode of human knowing and therefore likewise mistaken. One must be able to see a stone prior to knowing the stone and in knowing the intelligibility of 'stone' independent of the particular aspects of this or that stone, one knows the universal.

Analogy explains how Kant understood the notion of substance to be synthetic *a priori* and argues that the *a priori* nature of synthetic judgments is grounded for Kant *in the nature of human sensibility, our form of intuiting*. In the case of substantial change, Allison explains the transcendental condition: "the assignment of successively represented states of affairs to an enduring substratum functions as the rule through which we think of change. This assignment of successive represented states of affairs to an *enduring* substratum can be described as the form of the thought of a particular substantial change in the sense that to think such a change (as an object of possible experience) is just to connect one's perceptions according to the rule. This, in turn, makes it into a transcendental condition of the experience (experience is knowledge by means of connected perceptions) of a given substantial change."[150] This connection of an object of possible experience to an enduring substratum functions as a rule through which one thinks of substantial change.

For Kant the *manifold of appearances* are not of things in themselves, so synthetic *a priori* concepts are not of things as they are external to a knowing subject but rather the *empirical content of sensibility* is the *manifold of intuition* preconditioned by the *a priori* nature of both the *categories* and the *forms of sensibility* which are a diversity of appearances in a single intuition taken *a priori*. It follows that what can be known in the transcendental philosophy of Kant is nominal extension but never the comprehension or essence of things in themselves. Therefore for Kant what is known in cognition are synthetic *a priori* judgments that reunite in the *synthetic unity of apperception* objects (e.g., empirical objects such as a table or chair and *ens rationis* such as genus or species, etc.) preconditioned by the categories and the forms of sensibility. For example, 'all bodies are extended' or 'whatever begins to exist, proceeds from some cause' are both *a priori* propositions. Kant would argue that since the predicate is contained in the notion of the subject in the first instance, this proposition is analytic and because it is analytic it

---

150 Henry E. Allison, *Kant's Transcendental Idealism* (New Haven: Yale University Press, 1983), 206.

is independent of experience and is therefore *a priori* in that it is both universal and necessary being independent of experience.

In the latter case, the proposition fails to be *a posteriori* because the predicate cannot be united to the subject from experience. The causal maxim cannot be attained from experience based upon Hume's critique according to Kant. However, since the predicate is not contained in the subject, the *causal maxim* proposition can be synthetic. Moreover, because the predicate is reunited to the subject in the *synthetic unity of apperception* conditioned by the causal category and the *forms of sensibility* which are both *a priori*, universal and necessary, the latter proposition is not only synthetic but it is also *a priori*. Kant would classify the latter *causal maxim* proposition as a *schema* judgment. As Allison explains in reference to schema judgments in general:

> The heterogeneity of the intellectual and the sensible, taken together with the status of the transcendental schema as pure intuitions or conditions of sensibility, rules out the possibility that the claims are analytic, while the possibility that they are stipulative, which would make the connection arbitrary, is incompatible with Kant's claim that these schemata are the *sole* conditions under which the categories gain significance. Equally clear, a schema judgment cannot be synthetic *a posteriori*. This would imply that the connection between *category* and *schema* is based on experience; but this is incompatible with the *a priori* status of both, as well as with Kant's claim that it is only in virtue of these schemata that the categories can relate to experience in the first place. We are left, therefore, with the alternative that a schema judgment is both synthetic and *a priori*.[151]

The *transcendental schemata* is not only the *sensible conditions* that give meaning to and restrict the scope of pure concepts, they are also conditions that determine appearances in time, and thus make possible

---

151 Henry E. Allison, *Kant's Transcendental Idealism* (New Haven: Yale University Press, 1983), 186.

the possibility of experience. And it is because the *schemata* has this dual feature that they can mediate between appearances and pure concepts.[152] The second condition of the *schemata* that of being the conditions that determines appearances in time is discussed in relation to the *Analogy* and the *Principles of Pure Understanding*.

The Principles of Pure Understanding are those principles that are synthetic *a priori* judgments that assert that a particular *schema* functions as a necessary condition of the possibility of experience. For example, the principle of the *Axioms of Intuition* and *Anticipation of Perception* suggest that every intuition has extension and is therefore quantitative, and every sensation has intensive magnitude and thus it has degree or quality. Both quantity and quality are considered schemas or forms of sensibility.

The principle of the *Analogies of Experience* likewise claims that the schema of the relational categories of before, after, simultaneously functions as a condition for determining empirical time. Again, notice in each case, it is not quantity or quality or empirical relations of past, present or future that determine sensation and thus it is not what is known from experience that informs judgment, but rather the principles and schemata that determine what is sensed and known are that which inform transcendental judgment for Kant. Kant writes: "In the principle itself we do indeed make use of the category, but in applying it to appearances we substitute for it its schema as the key to its employment, or rather set it alongside the category, as its restricting condition, and as being what may be called its formula."[153]

It is also necessary to discuss the analogy and its relationship to the categories, the *schema*, and appearances. The *analogy* enables one to determine *a priori* for a given event y, there must be some antecedent event x from which y follows according to a rule and this rule is an analogy. Kant explains, "the relation yields a rule for thinking the fourth member in experience, and a mark whereby it can be

152 Ibid., 195.
153 *CPR*, A181/B224.

detected."[154] Thus the *analogy* provides a rule or decision procedure as a regulative routine or algorithm for finding a given term in experience. The *analogy* between pure concepts and their *schemata* (i.e., the forms of sensible intuition) enables the *schemata* to provide "translations into temporal terms of what is thought in pure concepts."[155]

Thus one finds that the *Analogies of Experience* act as a procedure that regulate or set the limits of sequential events in the temporal order determined by the relational categories and the schemata of time or causality such that a given causal event can only be before, after, or simultaneous. It is this determination that makes a given causal relation between cause x to be antecedent to event y in time (before, after, or simultaneous) and thus *a priori*. The principles such as the *Analogies of Experience* make use of the schemata, and *synthetic judgments* are *a priori* because the principles subsume appearances under the principles and these principles use the schema to condition appearances under *a priori* conditions. Thus the regulative nature of the analogies acting as procedures or as algorithms of apperception condition a given synthetic experience or intuition and thereby determine that experience to be *a priori*.

However, a causal event is determined by the *Analogies of Experience* to a certain time order, but it does not regulate causal succession itself apart from the pure concept of causality. The schema that orders in causal succession is the pure concept of causality, which is a rule-governed succession, or "the succession of the manifold insofar as it is subject to a rule." When Self in the unity of consciousness subsumes or subjects perception under the schema of causality that is under rule-governed succession, one consciously regards perception as containing the representation of an event, and as an object, the event itself is subsumed under the schema. The perceptions are not made into objects, but rather one conceives an objective temporal order through these perceptions where this objective order is thought in accordance

---

154 *CPR*, A180/B222.

155 Henry E. Allison, *Kant's Transcendental Idealism* (New Haven: Yale University Press, 1983), 196.

with the rule, that is, the rule is applied to the objective order and determines the conditions of perceptions as *a priori*.

In applying the given rule to the objective order determining the conditions of perceptions *a priori* Kant reverses the natural order of human knowledge. The following chapter will describe the mind-body relationship in Aquinas, which must be addressed prior to explaining how synthetic *a priori* knowledge of efficient causality is possible independent of the Kantian transcendental philosophy. Independent of the Kantian transcendental in the sense of being independent of the rules for the imaginative synthesis of the manifold or simply independent of the rules or schemata some of which are empirical and others *a priori*.

The mind-body relationship in Aquinas must be discussed in light of both early modern and contemporary developments and as well as the Cartesian divide between intelligibility and experience. The following chapters and two other supporting chapters not included reconcile how later developments of Galenic physiology by the Arab physicians, as a precursor to contemporary cognitive psychology, and the critique of various neo-platonist views of cognition during the Middle Ages successfully address how one can attain universality and necessity from experience, and how one can have synthetic *a priori* knowledge of efficient causality from experience independent of the transcendental philosophy. Independent of the transcendental philosophy in the sense of being independent of the rules for the imaginative synthesis of the manifold or simply independent of the rules or schemata.

Chapter 4

THE MIND-BODY RELATIONSHIP IN AQUINAS

This chapter will discuss in detail the psychology of Aquinas and its historical developments and its relationship to the mind-body synthesis achieved during the Middle Ages. In two chapters not included, the historical development of the internal senses are discussed in detail with an emphasis upon the cogitative power giving a basis for Aquinas's cognitive psychology in relation to contemporary cognitive psychology and physiology demonstrating a clear continuity between the former and latter. The chapter that follows will discuss further the relationship between the incidental sensibles and efficient causality, necessity, and universality.

Franz Brentano in *Nous Poiētikos* raised numerous questions concerning both medieval and more recent interpreters of Aristotle's conception of *nous* suggesting that separating the *nous poiētikos* from the human individual engender a confused misreading of Aristotle's psychology. Contrary to Brentano, a cogent argument will be developed to suggest that far from separating the *nous* from the human individual, Aquinas maintains a correct interpretation of Aristotle and frames the mind-body relation in such a way as to resolve many of the difficulties introduced by the Cartesian mind-body problem and the later critique of Hume indicating that the Kantian transcendental philosophy in fact addresses a false problem. In so arguing, this chapter and the next will develop a viable alternative consistent with contemporary physiology and psychology of perception and the Aristotelian tradition.

Brentano argues that "variations of the view that separates the *nous poiētikos* from the human individual ... discloses the contradictions necessarily imported into the Aristotelian doctrine by any such attempt, engendering a confusion that becomes greater the more one attends to particular passages."[156] For Brentano the Medieval conceptions of the agent intellect were problematic for this reason and Brentano's position seems to lean in the direction of simply doing away with the agent intellect. Brentano describes Avicenna's notion of two intellects described in Aristotle's *DA* 3.5 indicating that only the "intellect that becomes all things" (*intellectus materialis*) is found in man.[157] Avicenna following Alexander Aphrodisiensis held that the active intellect (*intelligentia agens*), the intellect that produces all things, was a purely spiritual substance, separate from man's nature.[158] Brentano explains that Alexander had held that "*nous poiētikos* is a purely spiritual substance, separate from the nature of man and acting upon him, the first ground of all things, the divine intelligence itself."[159]

For Avicenna the 'material intellect' (*intellectus materialis*) describes the Aristotelian potential intellect [*nous dunamei*] also following Alexander Aphrodisiensis.[160] The material intellect does not signify something corporeal although it is a cognitive faculty, but this faculty is a "passive substratum of the ideas, a capacity for thought" that apprehends intelligible forms.[161] The material intellect knows only potentially, but "for it to know actually, ideas must be imparted to it from some other substance, which is purely intellectual and separate from human nature."[162] It is this active intellect that "imparts the form"

---

156 Franz Brentano, "Nous Poiētikos" in *Essays on Aristotle's De Anima* (Oxford: Clarendon Press, 1992), 315.
157 Ibid., 315.
158 Ibid., 315.
159 Ibid., 314.
160 Ibid., 315.
161 Ibid., 315.
162 Ibid., 315.

such that "the images are capable of nothing but preparing the material intellect for the reception of the emanation."[163]

According to Brentano in Avicenna's psychology the "material intellect is illuminated by the light of the active intelligence and recognizes the general only if it looks upon the particular representations, which are in the imagination.[164] The activities of the imagination and of the sensory thought-faculty (*virtus cogitativa*) are needed to put it in a position to combine with the active intelligence and receive the intelligible forms that emanate from the latter."[165] After receiving the emanation of forms from the active intellect, they can only be retained in the imagination which is the "storehouse for the common sense or the fantasy."[166] Memory is the "storehouse for the estimative and sensory thought-faculty (storehouse of the intentions)."[167] Apprehension must return to these whenever representations are to be renewed.[168] Brentano concludes that "the intelligible form flows anew from the active intelligence into our material intellect so that learning is nothing but the acquisition of a perfect ability to combine with the active intelligence in order to receive the intelligible forms."[169] For Brentano this is a "strange transformation" of Aristotle's doctrine. Brentano concludes: "The sensory ceases to be the source of intellectual cognition; in a manner evidently approaching Plato, sensory representations are to constitute merely an occasion for our intellectual knowledge."[170]

---

163 Ibid., 315.

164 Brentano describes Avicenna's enumeration of the powers of the soul as follows: "(a) Sensory powers include (i) common sense or sensory representation (*sensus communis* or *seu phantasia*); (ii) imagination; (iii) the capacity for sensual judgment [*vis existimationis*; Avicenna takes this power to be located at the back of the precentral gyrus of the brain. It perceives the non-sensory meaning, which is contained in the sensible particulars; for example, in the sheep, it is the capacity of perceiving that this wolf is to be avoided while its own lamb is something to be cared for (elsewhere he calls it *aestimativa*, and in the case of man *cogitativa*); (iv) memory and reminiscence. (b) Intellectual powers: the capacity of action, which is the principle that moves the body toward action; the capacity of knowledge" (see Franz Brentano, "Nous Poiētikos" in *Essays on Aristotle's De Anima* (Oxford: Clarendon Press, 1992), nt. 31, p. 331).

165 Franz Brentano, "Nous Poiētikos" in *Essays on Aristotle's De Anima* (Oxford: Clarendon Press, 1992), 315.

166 Ibid., 316.

167 Ibid. 316.

168 Ibid., 316.

169 Ibid., 316.

170 Ibid., 316.

Brentano then discusses Averroës explaining that Averroës takes the discussion of *DA* 3.5 to hold that the two intellectual substances are distinct from sensation. Both the material intellect (*intellectus materialis*) and the active intellect (*intellectus agens*) are two immaterial substances that merge in intellectual knowledge and have their "nature separate from the human body as well as from each other."[171] On the other hand, for Averroës the passible intellect (*intellectus passibilis*) is a "sensory power located in the middlemost cell of the brain.[172] Through this faculty we distinguish individual representations and compare them with each other."[173] This is equivalent to Avicenna's *virtus cogitativa* and *virtus aestimativa* or *existimativa*. Averroës writes: "And it is through that intellect [which Aristotle calls passible *DA* 430a24] that man differs from other animals."[174]

The activity of the passible intellect is connected to the activities of the imagination and of memory in acquiring habitual knowledge. The habitual knowledge is not immaterial, but is the passible intellect. The passible intellect is corruptible and will perish. However, the material intellect is "merely the capacity for intelligible forms, and the active intellect makes actually intelligible the sensory pictures in man that are potentially intelligible (the images), and which thus moves the material intellect."[175] It is the material intellect that takes the concepts which are intelligible and located in the images, but the active intellect neither receives the images nor does the active intellect have knowledge of the images. The active intellect however makes knowable to the material intellect those concepts and images taken up as intelligible by the material intellect. However, for Averroës "the material intellect is one and the same in all individuals"[176]

---

171 Ibid. 317.

172 Both the Arab philosophers and Aquinas used the contemporary physiology of their day in opposition to the physiology of Aristotle. In this case, it was the Galenic theory of perception modified by Avicenna and Galen's physiology that Aquinas refers to when mentioning the Physicians. This will be discussed in much more detail below.

173 Franz Brentano, "Nous Poiētikos" in *Essays on Aristotle's De Anima* (Oxford: Clarendon Press, 1992), 316.

174 Ibid., 334.

175 Ibid., 317.

176 Ibid., 335.

Averroës held that their "real essence the intelligible forms are in the images. At first the faculty of sensory thought (the passible intellect), in combination with imagination and memory, prepares the images to receive the influence of the active intellect; through it they become actually intelligible. Once the active intellect has made the images intelligible, the material intellect, which stands to all intelligible forms in the relation of potentiality, receives from the images the concepts of sensible things. The intelligible form unites the image and the material intellect and by means of the image a form of the material intellect is united with us. It is through this material intellect that everyone comes to know the same things for they all share in this material intellect.[177] But since the images through which the material intellect is connected to each person are different, each person is united to the material intellect in a different way so each person does not know what the other person knows.[178] Each person share in common knowledge about the same kinds of things but each person does not know the same thing that each person knows.[179] The material intellect receives intelligible forms from all persons living and the intelligible forms are eternal and always new. They are eternal due to the material intellect and new due to the images.[180] Brentano's Aristotle had never dreamed of such an account and Aquinas fought against this "profound misinterpretation" with all his power.[181]

It is in this milieu that Aquinas must find a reasonable synthesis and avoid misinterpreting Aristotle. On the one hand was the neoplatonism of Avicenna influenced by Al-Fārābi and Alexander of Hales. On the other hand were the Latin neoplatonists who continued to maintain a mind-body dualism and Averroës who held a "profound misinterpretation" according to Brentano. Brentano seems to agree most with the general account given by Theophrastus the immediate

---

177 Ibid., 317-318.
178 Ibid., 317-318.
179 Ibid., 318.
180 Ibid., 318.
181 Ibid., 319.

successor of Aristotle as handed down by Themistius. This includes the view that Aristotle held that the active and the receptive intellect, which becomes all intelligible things, is immaterial.[182] Second, both intellects were taken as to capacities of one and the same subject.[183] Last, this subject was taken to be an essential constituent of man.[184] Brentano explains with regard to Aquinas: "his explanation coincides to a remarkable extent in all the above-stated points with the fragment of Theophrastus…"[185]

Brentano proceeds to describe the account given by Aquinas and raises a number of objections. Brentano explains that Aquinas not only takes the active intellect [*intellectus agens*] to be immaterial but also the potential (i.e., the possible) intellect [*intellectus possibilis*] and this in contrast to the Arab philosophers.[186] Second, Aquinas takes both the potential (possible) and the active intellect to belong to human nature as powers through which the soul acts and they are taken to be faculties of the soul. Brentano explains, that for Aristotle and Aquinas to say that the active and potential (possible) intellect are separate from the body only means that the potential (possible) and active intellect do not have organs like memory, the cogitative power (passive intellect or particular reason), and the phantasm which are faculties of the sensitive part.

Brentano notes that the "human soul stands on the boundary between the world of bodies and intellects… the human soul possesses powers that are not faculties of the ensouled body but belong exclusively to it; there remain for it some activities in which corporeal matter does not participate. In this way the potential and the active intellect are incorporeal in their activity and existence, unmixed with matter."[187] The potential (possible) intellect is the proper cognitive faculty but initially it is merely the capacity for thought like an empty tablet and

---

182 Ibid., 313.
183 Ibid., 313.
184 Ibid., 313.
185 Ibid., 319.
186 Ibid., 319-320.
187 Ibid., 320.

for this reason it is called the potential or possible intellect.[188] Brentano maintains that Aristotle held that the origin of our knowledge is in the senses and the soul cognizes nothing but images. However, since no corporeal thing can give an impression in something incorporeal, Aristotle proposes the active intellect. Since the power of sensory bodies is not sufficient to attain intellectual knowledge a higher agent is needed as an intellectual faculty of the soul for Aquinas. Brentano explains:

> The images which are received from the senses are only potentially intelligible since particular matter still adheres to them. The active intellect makes them actually intelligible through abstraction and for this reason it is the proper and preeminent (active) cause of intellectual knowledge, while the images are only the accompanying cause and, in a sense, the matter of the cause. The active intellect illuminates the images and abstracts the intelligible species from the images. It illuminates [*erleuchtet*] them; the images are to the intellect as colours are to the sense of sight. The influence of the active intellect prepares the images so that intellectual concepts can be abstracted from them, just as the sensitive part is raised to a higher power through its union with the intellective part. The active intellect abstracts the intelligible species from the images, i.e. through the power of the active intellect we can

---

188 Aquinas agrees with Aristotle: "the Philosopher compares the possible intellect to a blank tablet on which nothing is written (*De Anima*, III, 4, 430a1). Having more potentiality than other intellectual substances, the human soul is so close to matter that a material reality is induced to share its own being, so that from soul and body there results one being in the one composite, though this being, as belonging to the soul, does not depend on the body" (*On Being and Essence* Chpt. 4, Par. 10). Again Aquinas explains: "On the contrary, The human soul is naturally "like a blank tablet on which nothing is written," as the Philosopher says (De Anima iii, 4). But the nature of the soul is the same now as it would have been in the state of innocence. Therefore the souls of children would have been without knowledge at birth.

I answer that, As above stated (*ST* I, q. 99, a. 1), as regards belief in matters which are above nature, we rely on authority alone; and so, when authority is wanting, we must be guided by the ordinary course of nature. Now it is natural for man to acquire knowledge through the senses, as above explained (*ST* I, q. 55, a. 2; *ST* I, q. 84, a. 6); and for this reason is the soul united to the body, that it needs it for its proper operation; and this would not be so if the soul were endowed at birth with knowledge not acquired through the sensitive powers. We must conclude then, that, in the state of innocence, children would not have been born with perfect knowledge; but in course of time they would have acquired knowledge without difficulty by discovery or learning" (*ST* I, q. 101, a. 1, body; emphasis added).

grasp and consider the general nature of things without their individual determinations; the representations of this nature are received into the potential intellect as forms.[189]

Brentano indicates that this account differs entirely from that of the Arabs and that of Alexander but it does agree with Aristotle's immediate successor Theophrastus. Brentano confirms the consistency between Aristotle and Aquinas. However, Brentano advances a number of objections that raise the issue of the mind-body problem. First, Brentano is unable to reconcile how the active intellect can act upon the images or phantasms to generate thoughts in the intellect since a corporeal organ cannot act upon the incorporeal. However, Brentano neither discusses the role of the possible intellect in detail nor that of the cogitative sense, nor the instrumental causal relationship between the intellect and sense perception.

Second, Brentano argues that even if the intellect could transform the images into something intellectual, the intelligible species would not be the same after the transformation as before; they would no longer be images and Aristotle says that we can never think without at the same time having the corresponding image within us. Thus Brentano assumes or infers from this that Aristotle must not have held that the image is transformed into a higher or more intelligible thing at the moment of cognition. However, Brentano does not discussion the process of abstraction in any detail nor the need to return to phantasms or sense perception to attain sensible images.

Third, Brentano argues that for Aristotle a mere stimulation of the senses is not sufficient for knowledge acquisition since the intellect is devoid of ideas. For Plato mere sensible stimulation allows for recollection of forms, but for Aristotle the active intellect has no thought in it so how can it then be in a position to impart concepts to the potential intellect? However, Brentano fails to discuss in any detail the role of the active intellect in the process of abstraction and why the active intellect does

---

189 Franz Brentano, "Nous Poiētikos" in *Essays on Aristotle's De Anima* (Oxford: Clarendon Press, 1992), 320-321.

not convey concepts as Avicenna would have it. Thus the active agent intellect is that by which the possible intellect becomes all things, the intellect and the object are thus assimilated.[190]

Fourth, if one were to suppose, however, that the active intellect had the power to generate ideas when stimulated by a sensory representation, then the ideas would be potentially contained in the active intellect from the beginning but then this is exactly what Aristotle was attempting to avoid in Plato's account. However, again, this assumes an account of abstraction that imposes the need for such ideas to be contained in the

---

190 Of course, it should be mentioned that Richard Sorabji is correct to see an apparent "problem" in Aristotle's identity theory between the intellect and the intelligible. Theophrastus had asked which is the agent, the intelligible or the active intellect. Sorabji indicates that Alexander replied it is only "qua intelligible that the active intellect can be described as agent." Sorabji agrees with the commentators that "intelligibles are best seen as efficient causes." (see Richard Sorabji, "Aristotle on Sensory Processes and Intentionality. A Reply to Miles Burnyeat" in *Ancient and Medieval Theories of Intentionality*, Dominik Perler (eds.), (Brill, Leiden: 2001), 60). Similarly, Klubertanz indicates: "the discursive power is explicitly mentioned as a necessary link [for Aquinas]. Though the intellect can be a moving cause of our actions, it is always a remote (or principal) cause, while the proximate (and therefore instrumental) cause is the discursive power and the phantasm" (George P. Klubertanz, S.J., *The Discursive Power* (Ohio: The Messenger Press, 1952), 178).

The active agent intellect is the principle cause or the efficient causal agent (or, instrumental cause) as that by which intelligibles are apprehended in the process of abstraction from phantasms. The possible intellect is passive and the agent intellect is an active principle by which the possible intellect apprehends the intelligible species from sensible conditions. Similar to light making possible abstraction of color in the visual operation of abstracting color from matter in transient motion, the light of the agent intellect makes possible the apprehension of intelligibles through a process of abstraction from material conditions in immanent motion. The internal faculties are instrumental and material causes, at least in preparing sensible forms for human cognition, since the proper mode of knowing is from the sensible. The difficulty indicated by Sorabji may be due to a misunderstanding of transient and immanent motion. Aquinas commenting on Aristotle explains that one transient motion "insofar as it proceeds from the mover to the mobile object, is the act of the mover. But this same motion, insofar as it is in the mobile object as coming from the mover, is the act of the mobile object" (*In Phys.* III, 4.10-11 (#306-307) All citations from St. Thomas Aquinas *Commentary on Aristotle's Physics*, Translation by R. J. Blackwell et. al. New Haven: Yale, 1963 unless otherwise noted). Taken in terms of transient motion, insofar as this motion is the act of the agent or mover, this motion is "the act of the active" and is called "action." Insofar as this motion is the act of the mobile object, it is the "act of the passive," and called "passion" (*In Phys.* III, 5.2 (#309)). However, if taken as immanent motion, the operation of the intellect is understood as the "act of something existing in act." (*ST* I, q. 18, a. 3, ad 1) and is different from transient motion. In the case of transient motion, action proceeds from the agent to the mobile object. In the case of immanent action, the operation (immanent motion) remains in the agent. This is one distinction between transient and immanent motion: in the case of immanent motion, action or operation remains in the agent. Thus in the intellectual operation, the operation remains in the agent. The possible intellect becomes all things in the act of the agent intellect making all things. In addition, in the case of transient motion, action is the perfection of the object. However, in the case of immanent action, operation is the perfection of the agent itself. Hence, in the intellect knowing something, the operation of knowing is the perfection of the intellect. Therefore the intellectual operation is an efficient cause and the intellect is that which is perfected in the operation of knowing.

active intellect; however, this is not the case if the process of abstraction does not require the active intellect to contain such ideas. Abstraction for Aquinas is from sensibility to intelligibility and not from the agent intellect to the possible intellect. Nor is it from the transcendental as Kant would have it. Brentano, like "transcendental Thomism," seems to have misunderstood Aquinas in this regard.[191]

Fifth, in this case, it is inexplicable why a lack of sensation should always lead to a lack of cognition, or why cognition is possible only so long as one retains the corresponding particular representation in the imagination if the ideas are already potentially contained in the active intellect. Again, Brentano seems to interpret Aquinas according to a neo-Platonist influenced theory of emanation where ideas are already potentially contained in the active intellect and received by the intellect in a manner similar to the doctrine of Avicenna.[192]

191 Brentano fails to explain in his discussion of Aquinas the relation between sensible forms and first intention intelligible forms, or the connection between the corporeal internal senses and the incorporeal intellect in Aquinas. Nor does Brentano discuss the relationship between the potential (possible) intellect and the passible (cogitative power) intellect in Aquinas. The passible intellect is the cogitative faculty (*vis cogitativa*) where sense inquiry about material singulars takes place and the cogitative internal sense acts as a kind of bridge between the corporeal sense and the incorporeal intellect. The intellect knows singulars only indirectly by returning to phantasms as it were by a kind of reflection in turning to the phantasms through the cogitative power and abstracting the intelligible species from individual matter. The cogitative power being a corporeal faculty knows the singulars directly through the phantasms and sensible memory. Thus the intellect "understands the universal directly through the intelligible species, and indirectly, the singular represented by the phantasm" (*ST* I, q. 86, a. 1). The agent and possible intellect are responsible for the intelligible universal as abstracted from the singular and the cogitative power along with the other internal senses are responsible for the sensible singular. Aquinas following Averroës defined more precisely the faculties required for cognition but maintained with Avicenna that these faculties of the internal senses are to be found in Aristotle. Interestingly, Hans Meyer explains that the *Comment. On the Sentences.*, III, 33, 2 ad 4 describes the location of what Aquinas called the *ratio particularis*, i.e., the *intellectus passivus* or the cogitative power which is "completely dependent on the body, having the pineal gland as its organ." (cf. Hans Meyer, *The Philosophy of St. Thomas Aquinas* (St. Louis: B. Herder Book Co., 1944), 184). Aquinas seems to have followed the medical tradition in attributing the cogitative power to the pineal gland, and similarly Descartes mistakenly maintained that the pineal gland was the location of the *sensus commune* (common sense). Today it is understood that the pineal gland receives input from visual pathways through the hypothalamus where lighting conditions inhibit normal secretion of melatonin regulating sleep and can cause seasonal affective disorder (SAD): "When neurons of the retina are stimulated by light entering the eye and impulses are sent to the hypothalamus, these messages eventually reach the pineal gland where they inhibit secretion of melatonin" (Goodenough, Judith, Robert A. Wallace, Betty McGuire, *Human Biology* (Fort Worth: Saunders College Publishing; Harcourt Brace College Publishers, 1998), 101, 158).
192 Aquinas explains: "Avicenna holds (Avicenna, *Metaph.*, IX, 4-5 (105r)) that just as sensible forms are not received into sensible matter except through the influence of the agent intelligence, so, too, intelligible forms are not imprinted on human minds except by the agent intelligence, which for him is

Brentano continues this line of questioning attempting to develop as many contradictions as possible to refute the account of Aristotle, Theophrastus, and Aquinas along the lines of Avicenna and Averroës. One can infer from Brentano's criticism that he fails to see how Aquinas's account can possibly bridge the impassible chasm between sensibility and intelligibility introduced by early modern philosophy. The essential question to be addressed is how can the active intellect act upon the images or phantasms to generate thoughts in the intellect since a corporeal organ cannot act upon the incorporeal?

It will be necessary to discuss Aquinas's development of the hylomorphic view of the mind-body relationship specific to the cogitative power and the process of abstraction in response to Brentano's critique. This will include intentionality, the process of abstraction, the relation between universals and particulars, sensible and intelligible forms, and the basic cognitive psychology of Aquinas. The following sections will discuss the process of abstraction in the first operations of the intellect, and discuss Aquinas's understanding of the internal senses and Aquinas's cognitive psychology of the cogitative sense and incidental sensibles related to causality and a natural knowledge of God in preparation for the chapter on efficient causality and modality.

Mind-Body and Substantial Unity in Aquinas

As noted by Brentano, the different operations of the sensitive power and those of the intellect advance the question, how can universals be abstracted from sensible forms?[193] A further question to be raised

---

not a part of the soul, but a separated substance. However, the soul needs the senses to prepare the way and stimulate it to knowledge, just as the lower agents prepare matter to receive form from the agent intelligence. But this opinion does not seem reasonable, because, according to it, there is no necessary interdependence of the human mind and the sensitive powers. The opposite seems quite clear both from the fact that, when a given sense is missing, we have no knowledge of its sensible objects, and from the fact that our mind cannot actually consider even those things which it knows habitually unless it forms some phantasms. Thus, an injury to the organ of imagination hinders consideration. Furthermore, the explanation just given does away with the proximate principles of things, inasmuch as all lower things would derive their intelligible and sensible forms immediately from a separated substance." (*De Veritate* II, q. 10, a. 6, body).

193 Alfred Leo White, Jr. explains the problem between sensibility and intelligibility expressed by

is how can perception of efficient causality be taken from sensibility and how can perception of efficient causality from sensibility lead to universal and necessary causality? To answer these questions one must turn to the mind-body synthesis developed by Aquinas. Following the seventh chapter of *De Divinis Nominibus* of Dionysius, Aquinas argued that every "inferior nature in its highest element touches the lowest element of a superior nature, according as it participates in the superior nature, although deficiently; therefore, in apprehension there is to be found something in which the sensitive part touches reason."[194] The essential question raised by Brentano's critique is how can the active intellect act upon the images or phantasms to generate thoughts in the intellect since a corporeal organ cannot act upon the incorporeal without the agent intellect having innate ideas or be a separate substance as Avicenna had maintained? It is here that one must turn to the technical term *continuatio* (or *coniunctio*) in Aquinas.

Among the Arab philosophers, *continuatio* suggested a "union between the separated intellect and man, in and through an operation."[195] St. Thomas rejected the notion of a separate substance but indicated instead a union between intellect and sensibility through operation or causality.[196] For Aquinas the human intellect comes into being as mere potency and therefore requires senses and the act of sensation requires a body.[197] Therefore, the human intellect requires a composite

---

Brentano: "both the sensible object and the sense power have the same, material mode of being. Hence the sensible object is already suited to immute the sense power. The object represented by the phantasm, however, is material; hence it cannot impress the intellect, for nothing material can impress something immaterial. In order to be able to impress the intellect, the phantasm must somehow become actually immaterial or actually intelligible. The intellect therefore must somehow act upon the intellect in order to make it actually intelligible" (Alfred Leo White, *The Experience of Individual Objects In Aquinas A Dissertation* (Washington, D.C.: The Catholic University of America, 1997), 226).

194 *In III Sent.* d. 26, q. 1, a. 2, sol., ed. Pierre Mandonnet, O.P., and M. F. Moos, O.P., Paris: Lethielleux, 1929-1933, vol. III, pp. 816-17 in George P. Klubertanz, S.J., *The Discursive Power* (Ohio: The Messenger Press, 1952), 153.

195 George P. Klubertanz, S.J., *The Discursive Power* (Ohio: The Messenger Press, 1952), 167.

196 Ibid., 167.

197 Aquinas describes this relationship in a text opposed to Platonic and Augustinian dualism: "A human soul does not possess innate intelligible species by means of which it can accomplish its essential operation, which is to understand, as higher intellectual substances do; but rather a human soul is in potency to intelligible species since it is like a wax tablet on which nothing has been written, as is said in Book II of the *De Anima* (Ari., *De Anima*, III, 4 (429b 30-430a 1)). Consequently it must acquire

being such that intelligibility and sensibility are in fact related as two powers of one composite being, and the intellectual power consists both of the cogitative faculty and the intellect.[198] The *possible* (i.e., the potential) *intellect* is separate and unmixed while the *passive intellect* or the *cogitative* faculty has a corporeal organ. The act of intellectual apprehension and perception are united in act to the sensible power of the cogitative faculty, memory, and the *phantasmata* in a composite operation where the intellect acts instrumentally upon the corporeal faculty in a cognitive operation of the particular.[199]

For Aquinas, this soul-body composite is a real unity as form is to matter and the indirect intellectual knowledge in returning to sensibility in apprehending knowledge of the singular is one discursive power.[200] The bridge between sensibility and intelligibility is the highest sensitive power in one faculty touching the lowest intellectual power in the same faculty, and this faculty is the cogitative power or particular reason

---

intelligible species from things outside itself through the mediation of sense powers, which cannot accomplish their appropriate operations without bodily organs. Hence it is necessary that a human soul be united to a body" (*Quastio de Anima*, q. 8 response). Similarly, "Nec tamen est verum quod aliquis actus sit hominis in vita praesenti in quo corpus non communicet; quia, quamvis in actibus intellectivae partis non communicet corpus sicut instrumentum actus, communicat tamen sicut repraesentans obiectum; quia obiectum intellectus est phantasma, sicut color visus, ut dicitur in III *De Anima*. Phantasma autem non est sine organo corporali; et sic patet quod etiam intelligendo, et in aliis actibus animae, utimur aliquo modo corpore," (Cf. George P. Klubertanz, S.J., *The Discursive Power* (Cincinnati, Ohio: The Messenger Press, 1952), 169).

198 There have been numerous theories proposed to address the mind-body interaction including Cartesian interactionism or the pineal gland theory, the corpus callosum theory refuted by Zinn's split-brain experiments, the occasionalism of Malebranche where there is no interaction just divine synchronization, the parallelism of Leibniz where no interaction exists -- there is simply a *hormonia praestabilita*, the monism of Spinoza where there is one substance (*deus sive natura*), of course there is the materialist account of Lamettrie (1709-1751) holding that "all mental processes are completely determined by physical processes" and that of Francis Crick in *The Astonishing Hypothesis: The Scientific Search for the Soul* (1994), also the Turing Machine functionalist theory of Putnam and Hilary, and of course numerous mathematical or computational models of the mind-brain interaction (for an account of these theories, their development and a critique, see Gert-Jan C. Lokhorst "Philosophy and the Brain." Paper delivered at the University of Helsinki, May 14-15 2001 (transparencies available at http://www2.eur.nl/fw/staff/lokhorst/helsinki.2001.html)). For an assessment of the various proposed models including functionalism and Postulational Realism along with the Hylomorphic alternative, see William Jaworski, "Hylomorphism and the Mind-Body Problem." Paper delivered at the American Catholic Philosophical Association, November 5-7, 2004.

199 For a textual study of St. Thomas's view of *continuatio* and knowledge of the singular one can see G. P. Klubertanz, S.J., "St. Thomas and the Knowledge of the Singular," *The New Scholasticism*, XXVI (1952), 135-63.

200   *In De Anima* I, 10.152-156, 160-163.

deriving its excellence not from the sensitive part but from universal reason. Aquinas explains "the discursive power is that which is highest in the sensitive part, where the latter somehow touches the intellective part so as to participate that which is lowest in the intellective, namely, the discourse of reason, according to the rule of Dionysius, *De Divinis Nominibus*, c.7 that 'the beginnings of the second things are joined to the ends of the first.' And so the very discursive power is called 'particular reason,' as is clear from the Commentator in the third book of the *De Anima*, and is to be found only in man".[201]

It is by this discursive power, called particular reason also known as the passive intellect, acting as a sensitive faculty but participating in the intellectual act and joined to the intellect in a composite operation, that the phantasms are prepared for abstraction by the light of the agent intellect. The agent intellect reflects the images found in the phantasms through the cogitative power converting the material into the immaterial and sensibility into intelligibility whereby the potential intellect also known as the possible intellect becomes all things in receiving as its proper object the quiddity or what a thing is.[202] In this

---

201 *De Veritate*, XIV.1 ad.

202 Aquinas commenting on the mistaken position of Averroës that held that the particular reason rather than the possible intellect is that which distinguishes man from other animals explains the essential operation of the cogitative power: "The work of the discursive power is to distinguish individual intentions, and to compare them among themselves, just as the intellect which is separate and not mixed [with matter] compares and distinguishes universal intentions. Because through this power, together with the imaginative and the memorative, phantasms are prepared to receive the action of the agent intellect by which they become actually intelligible (just as there are some arts which prepare the matter for the principal artisan), so this power is called by the name of intellect and reason, and of it the physicians say that it has its seat in the middle cell of the head" (*SCG* II.60; All *SCG* citations are from St. Thomas Aquinas. *Summa Contra Gentiles*, trans. by English Dominicans, (London: Burns, Oates, and Washbourne, 1934) unless otherwise noted). Aquinas argues that Aristotle demonstrated that the principle of man's proper operation that distinguishes man must be impassible and not mixed with the body. That which gives man his specific nature is related to the act of understanding as first act to second act, thus the possible intellect which is unmixed and impassible is that which distinguishes man and not the passive intellect (*SCG* II, 60.2).

Aquinas also argues against Averroës that the habit of science cannot be of a corporeal body but must be an operation of the possible intellect (*SCG* II, 60.12). Further, a knower is assimilated to the thing known only with respect to universal species, and these universal species are the objects of science which cannot be a power of an organ, and therefore must be in the possible intellect (*SCG* II, 60.13). From this intellectual memory as a function of the possible intellect resides in the possible intellect and is the location of the *habitus* of science and of first principles (*SCG* II, 74.12). Aquinas rejects the notion of Avicenna who proposed that the agent intellect is a separate substance separate from the soul that

way, the indirect intellectual knowledge of the singular is one act or operation, involving the use of two distinct powers -- the intellect and sensitive powers. Franz explains:

> Any "continuation" or "participation" of the *vis cogitativa* in universal reason takes place through the agent intellect, not entitatively, but operatively. Entitatively the *vis cogitativa* remains a sense faculty and the agent intellect remains a spiritual faculty. However, *operatively* the *vis cogitativa* participates in the activity of the agent intellect and the agent intellect receives specification from the activity of the *vis cogitativa*. Therefore, the *vis cogitativa* acts not only through its own power but also through the action of the agent intellect.[203]

Again Aquinas explains: "The cogitative and memorative powers in man owe their excellence not to that which is proper to the sensitive part; but to a certain affinity and proximity to the universal reason, which, so to speak, overflows into them. Therefore they are not distinct powers, but the same, yet more perfect than in other animals."[204] This is a response to an objection that the cogitative power which consists in comparing, adding, and dividing is not essentially different from

---

infuses intelligible species into man's possible intellect (*SCG* II, 74.2,5). Aquinas says this is no different than the infused knowledge held by Plato and contrary to experience. Experience indicates that our knowledge is caused by sensible things, one who lacks one sense also lacks the knowledge of those sensible things which are known through that sense (*SCG* II, 74.5). The agent intellect causes things to be actually intelligible and the possible intellect is the recipient of those things that are intelligible (*SCG* II, 60.14). The intelligible species is the act of understanding and the possible intellect is the principle of this operation (*SCG* II, 62.10). The possible intellect is a certain act and *not* that which prepares the intelligible species since "preparedness results from a blending of the element" and "clearly depends on the body" (*SCG* II, 62.9, 12-13). Aquinas also indicates, "Aristotle says that the power of sense, which occupies the lowest place in the order of cognitive power, is 'receptive of sensible species without matter.'… the possible intellect is the highest cognitive power in us; for Aristotle says that the possible intellect is 'that by which the soul knows and understands'" (*SCG* II, 62.7; emphasis added). Also note that experience itself indicates that knowledge is caused by sensible things rather than the Kantian empirical or *a priori* schemas as illustrated by the fact that if one sense is not available, knowledge of those sensible things associated with that sense remain unknown.

203 Edward Quinlisk Franz, *The Thomistic Doctrine on the Possible Intellect* (Washington D.C.: CUA Press, 1950), 131. Likewise Aquinas writes: "For the impression of the superior agent remains in the inferior, and from this the inferior agent not only acts through its proper action but through the action of the superior agent." (*De Veritate* III, q. 22, a. 3c: "Impressio enim superioris agentis manet in inferiori, et ex hoc inferius agens non solum agit actioni propria sed actioni superioris agentis."

204 *ST* I q. 78, a. 4 reply 5.

the reminiscent power that consists in a kind of discursive reasoning process in recollection. Aquinas agrees that the aspect of the memorative power that acts discursively derives its affinity from the intellect and it is therefore in this regard not essentially a distinct power from that of the intellective power, and as such the memorative power because of syllogistic reminiscence is more perfect in humans than in other animals. In addressing the objection, Aquinas further reflects upon the relationship between the distinct powers of sensibility and intelligibility and explains that the cogitative and memorative faculties specific to discursive reasoning are one power in that they share in their affinity and proximity to the universal reason and owe their excellence to the intellectual power which overflow into them.

Aquinas develops the principle that powers of the soul are distinguished by the different formal aspects of their objects. The reason for this is that something is to be defined in terms of the object to which it is directed. Further, if there is a common *ratio* between various aspects of a given object, then the power by its nature will not be differentiated according to the individual differences of that object. For example, in the case of color it matters not that the color is green or yellow since color is the common *ratio* by which the object of color is known as the proper sensible of sight. Aquinas argues that the intellect regards its proper object under the common *ratio* of being because the possible intellect taken as that which is potential is potentially all things. Therefore the intellect is not to be distinguished by any differences of being. There is a distinction, however, between that which is passive and that which is active and therefore a distinction between the possible and active intellect does exist.

Aquinas argues, memory and the cogitative power like the possible intellect are potentially all things "for it belongs to the nature of the passive faculty to retain as well as to receive."[205] Since there can only be a distinction between the active and potential, there cannot be any other differences of powers in the intellect.[206] In addition, "the act

---

205 *ST* I q79, a7, body.
206 *ST* I q79, a7, body.

of reason is a movement from one thing to another," and the same movable being passes through the "lower reason to the higher reason." Hence, the two operations are the same intellective power.[207] Aquinas quotes Augustine, "the higher and lower reason are only distinct by their functions" and adds "therefore they are not two powers."[208] Therefore memory, the cogitative power, and the possible intellect are three functions of the same intellective power or operation insofar as discursive reason is involved, but although they are one power, they are distinct by their functions. If one looks at the movement from the two powers of sensibility and intelligibility, one sees two operations one of the lower and one of the higher reason. However, if one looks at the movement from the point of view of one intellective power, the movement from lower to higher is one intellective operation.

St. Thomas can hold therefore there are a number of powers where there is composition and contact between these various powers in virtue of a composite subject. Form is the formal determination of each faculty according to its participation in the higher and lower powers of a substantial composite being, and in this composition, Aquinas can maintain distinctions based upon function or operation. The key to this composition and unity of distinct powers is the Dionysian principle of hierarchy, an order of participation—"an order of extrinsic formal causality."[209] Klubertanz explains:

> If the body is really a part of man's being, then it likewise has an intelligible work to do, even in relation to man's proper and specific work, which is understanding. Again, if one admits that the human intellect is itself in potency, and receives its objects from sense through imagination, then the dependence of intellect in all its operations is decided in principle.... There was a tendency among philosophers to deny the real diversity and distinction of powers, and to insist on the unity of human operation. St. Thomas was quite ready to grant

207 *ST* I q79, a9, body.
208 *ST* I q. 79, a. 9, contra.
209 George P. Klubertanz, S.J., *The Discursive Power* (Ohio: The Messenger Press, 1952), 171.

that the act of perception presents itself as one—as a concrete unit (this doctrine comes clearly to the fore in the discussion of the "sensibles *per accidens*"; cf. *In Sent.* IV, d. 49, q. 2, a. 2; *In De Anima* II, lect. 13; *Summa Theologiae* I.12.3 ad 2). Nevertheless, he also maintained ... a discovered multiplicity of powers, and an experimentally given unity of operation. Such a situation calls for a synthesis that will unify without obliterating all distinctions.[210]

It is in this that the critique raised by Brentano is addressed not at the level of propositional semantics attributed to a functional psychology, but at the level of composite unity among distinct powers that share a common mode of operation. How does the active intellect act upon the images or phantasms to generate thoughts in the intellect since a corporeal organ cannot act upon the incorporeal? If in fact the intellect is incorporeal as Aquinas argued, the solution to Brentano's criticism is the composite substantial unity between the internal senses and the possible and active intellect as they relate to the hylomorphic view of the mind-body relationship following the Dionysian principle of hierarchy found in the Aristotelian tradition.

If the intellect is not incorporeal, then clearly there is no distinction between corporeal sensibility and incorporeal intelligibility. However, in such a case, one is left with the need to explain how material conditions can generate what appear to be immaterial cognitive acts such as universals and second intention concepts. Contemporary neuro-physiology and cognitive psychology cannot yet explain the immaterial nature of universals, nor that of intentions since material conditions do not in themselves generate the immateriality of universals, nor are abstract ideas merely material even if they are associated with bio-chemical and neurological activity as the dualist-materialist dichotomy would have it. Further, incidental intentions not specifically associated with a proper external sense cannot be explained by material causality entirely independent of cognition. Further, the various theories of

---

210 Ibid., 171.

perception and the irreducible complexity of microbiological neuro-physiology itself suggests that there are aspects of cognitive psychology not yet understood. Nor can a dualist-materialist approach explain what appears to be a synthetic unity between materiality and immateriality found in cognitive acts.[211]

Alfred Leo White likewise explains "in the *Summa theologiae* he [Aquinas] revised Avicenna's notion of lower immaterial principles flowing from higher ones, applying it to the relationship between powers of the soul. Lower powers flow from the soul through the mediation of higher powers and participate in the acts of the higher powers.[212] The cogitative and memorative powers therefore proceed from the soul via the power of reason, so that rationality 'flows' into these powers. The human cogitative and memorative powers can thereby perform operations that go beyond the estimative and memorative powers of brutes."[213] The Dionysian principle serves to tie the intellect to the sensible cognitive operations of the discursive power of cogitative perceptive judgment but also to the memorative powers in the act of discursive reminiscence where recollection and the process of association takes place in moving from prior knowledge to subsequent knowledge through a discursive process of association moving from the known to the unknown and "terminating in a new act of remembering."[214]

---

211 Alfred Leo White, Jr. emphasizes this point: "a higher sense power is able to appropriate the acts of lower powers precisely in virtue of sensation's being an operation (i.e., what Aristotle would call an *energeia*) rather than a sheer process or natural immutation. Only recourse to this metaphysical understanding of sensation can affirm the unity of our sentient awareness. A similar appropriation of the acts of lower powers seems to be involved in intellectual apprehension and in the subsequent reflection through which the intellect knows the individual. Those, on the other hand, who would reduce sensation to material processes are destined to fragment human consciousness into many separate activities, linked together by a kind of neural conveyer belt" (Alfred Leo White, Jr. *The Experience of Individual Objects In Aquinas A Dissertation* (Washington, D.C.: The Catholic University of America, 1997), 8).

212 Aquinas explains: "Therefore one power of the soul proceeds from the essence of the soul by the medium of another...those powers of the soul which precede the others, in the order of perfection and nature, are the principles of the others, after the manner of the end and active principle... The senses, moreover, are a certain imperfect participation of the intelligence..." (*ST* I, q. 77, a. 7, c. *Summa theologica*, trans. by the Fathers of the English Dominicans Province (London: 1920; 2d reprint, Westiminister, *Md.*: Christian Classics, 1981)).

213 Alfred Leo White, Jr. *The Experience of Individual Objects In Aquinas A Dissertation* (Washington, D.C.: The Catholic University of America, 1997), 115.

214 Ibid., 116. Aquinas indicates: "And this is reminiscing: when in some way we resume a prior

Similarly, the same Dionysian principle of hierarchy is used to explain the relationship between the *sensus communis* and the proper senses, "the common sense, being a higher power, is the principle and terminus of the proper senses. More precisely, the common sense power is the proximate source or active principle from which the lower, proper sense powers flow and its operation is the end of the operations of the proper senses. Secondly, the common sense, like the proper senses, receives form immaterially; but because it is a higher power it is able to receive form in a nobler, more immaterial manner than any of the proper senses. Thirdly, the power and operation of the proper senses are participations in the power and operations of the common sense."[215] The Dionysian principle serves to allow the common sense to be the active principle for the proper senses and that from which the proper senses flow.

The common sense is able to receive forms in a more immaterial manner than the proper senses and has discerning judgment being a higher power than the proper senses.[216] The common sense "uses the acts of the proper senses as a principal efficient cause uses instruments."[217] It is from this participation in the *vis cogitativa*, following the Dionysian principle of hierarchy, that the common sense "both performs the judgment attributed to the external senses (e.g., distinguishing white

---

apprehension, ... but memory happens through reminiscing, for reminiscing is a certain movement toward remembering, and thus memory follows reminiscing, as a terminus of movement." "... [E]t tunc est reminisci, scilicet cum aliquo modo resumimus priorem apprehensionem, ... set per reminiscenciam accidit memorari, quia reminiscencia est quidam motus ad memorandum, et sic memoria sequitur reminiscenciam, sicut terminus motum." *Sentencia libri De memoria et reminiscencia* [*Commentary on De Memoria et reminiscencia*], cap. 4 (45.2:118.110-22).

215 Alfred Leo White, Jr. *The Experience of Individual Objects In Aquinas A Dissertation* (Washington, D.C.: The Catholic University of America, 1997), 50.

216 Aquinas explains: "the discerning judgment must be assigned to the common sense; to which, as to a common term, all apprehensions of the senses must be referred: and by which again, all the intentions of the senses are perceived; as when someone sees that he sees. For this cannot be done by the proper sense, which only knows the form of the sensible by which it is immuted, in which immutation the action of sight is completed, and from which follows another [immutation] in the common sense which perceives the act of vision." *ST* I, q78, a4, ad2). "In a physical head there is not only the power of sensing, in order that it may sense by sight, hearing, and touch and such senses; but this [sensitive] power is in it in such a way that it is the root from which sensation flows into all the other members" (*De Veritate* III, q. 29, a. 5, body).

217 Alfred Leo White, Jr. *The Experience of Individual Objects In Aquinas A Dissertation* (Washington, D.C.: The Catholic University of America, 1997), 50.

from black), yet also performs a judgment proper to itself because it exceeds the ability of any of the proper senses (e.g., distinguishing white from sweet).[218] By introducing the notion of the common sense as a principal efficient cause, Aquinas implies that this sense appropriates rather than duplicates the acts of the proper senses. In this way he conveys how the proper and common senses operate in a unified manner in perceiving the sensible object."[219]

The proper sense senses an external sensible object, but the common sense perceives the operation of sensing the external sensible object; however, the common sense also distinguishes between the external sensible objects so it also apprehends the external objects. The operation of the common sense include distinguishing between the external sensible objects of the proper senses and perception of sensation itself or the operations of each of the proper senses. It is due to the Dionysian principle "everything is received only according to the mode of the receiver"[220] and "the known is in the knower" that the common sense "unites its operation with those of the proper senses by receiving their acts within itself, just as each proper sense receives the act of the sensible object."[221]

Likewise, by an imperfect participation in the common sense, the proper senses are able to distinguish between various colors.[222] As

---

218 In the literature of the period the *vis cogitativa* would be associated with the portion of the brain called the cerebral cortex in contemporary cognitive psychology and physiology. The *vis cogitative* for Aquinas participates in the cognitive power that includes the passive intellect (particular reason), possible intellect (potential intellect), and the active (agent) intellect.

219 Ibid., 50-51.

220 Dionysius' *Liber De causis* in *Scriptum super libros magistri Petri Lombardi*, vol. 1, d. 38, q. 1, a. 2, ed. R. P. Mandonnet [Paris: P. Lethielleux, 1929], 901.

221 Alfred Leo White, Jr. *The Experience of Individual Objects In Aquinas A Dissertation* (Washington, D.C.: The Catholic University of America, 1997), 55.

222 Ibid., 55-56. Aquinas denies that the common sensibles (e.g., number, shape, extension, movement) are proper objects of the common sense (White, 43). The proper objects of the common sense are the proper sensibles such as color, flavor, odor, etc. (White, 43). However, the individual through the common sense, perceives "not only proper sensibles, such as the visible and the audible, but also common sensibles, i.e., shape, size, number movement, and rest" (White, 45). The common senses are called common because of having more than one sense. The common senses are given through the external senses whose proper objects are the proper sensibles. In this way the proper object of the common sense are the proper sensibles and the common sense perceives through these proper sensibles the common sensibles (see White, 45-46). White explains that the "common sense performs a judgment,

Aquinas explains: "One power of the soul proceeds from the essence of the soul by the medium of another," and it is in this unity of conscious perception united with the neuro-cognitive operation and according to the mode of the receiver that the higher powers receive in themselves the act of the lower powers and the higher powers flow to the lower powers analogous to a principal of efficient causality using instruments,[223] and by this a mind-body hylomorphic relationship exists.[224] Aquinas's

---

through which it perceives how the objects of the various proper senses belong together to form the same whole discerning different genera of proper sensibles acting on the senses and discerning the common sensibles of those proper sensibles acting upon the external senses" (White, 189-190). However, this is not an integration or synthesis of diverse proper sensibles belonging to various individuals acting upon the senses.

Instead, from what was perceived separately by the common sense and imagination, the higher cogitative power is that which interrelates the objects of the lower internal senses as past, present, and future in one perceptual act. It is the sensible powers that are aware of time, space, and the individual rather than the transcendental has proposed by Kant. The cogitative power in perceiving "this man" or "this woman" upon seeing something colored, distinguishes one individual from another and integrates proper sensibles with each other (White, 196-198). For Aquinas the universal qua universal is not spatio-temporal since the universal is abstracted from all material conditions (see Ibid., 150; ST I, q. 16, a. 7, ad. 2; In Posterior Analytics II, 42; ST q. 85, a. 1; body and ad. 1). Aquinas attributed to the cogitative power the awareness of incidental sensibles as its proper object and perception of sensible qualities as belonging to an individual following Averroës since the discursive intellective power discerns or judges between proper, common, and incidental intentions relating them to the individual. One finds in Aquinas the power of sensation, an imperfect reflexive consciousness in perception related to the common sense, a synthetic manifold of sense impressions in the case of the phantasm, and a discursive power that discerns individual intentions as past, present, and future at the level of sensibility and these powers are located in the brain itself whose operations flow from the higher intellective power of the soul, and it is this intellective power that understands its own act of understanding in self-reflection and is therefore self-consciousness in the more proper sense.

223 Aquinas indicates that "what is moved by another has a twofold action---one which it has from its own form---the other, which it has inasmuch as it is moved by another; thus the operation of an axe of itself is to cleave; but inasmuch as it is moved by the craftsman, its operation is to make benches. Hence the operation which belongs to a thing by its form is proper to it, nor does it belong to the mover, except in so far as he makes use of this kind of thing for his work: thus to heat is the proper operation of fire, but not of a smith, except in so far as he makes use of fire for heating iron. But the operation which belongs to the thing, as moved by another, is not distinct from the operation of the mover; thus to make a bench is not the work of the axe independently of the workman. Hence, wheresoever the mover and the moved have different forms or operative faculties, there must the operation of the mover and the proper operation of the moved be distinct; although the moved shares in the operation of the mover, and the mover makes use of the operation of the moved, and, consequently, each acts in communion with the other" (ST III, q. 19, a. 1 body). The communion of each follows from the second action where the operation of the instrument is not a distinct operation from that of the mover, and although the mover and moved have different forms or operative faculties, the instrument shares in the operation of the efficient cause. Thus each acts in communion with the other providing unity. (see Alfred Leo White, Jr. The Experience of Individual Objects In Aquinas A Dissertation (Washington, D.C.: The Catholic University of America, 1997), 60-63.

224 Aquinas explains: "One power of the soul proceeds from the essence of the soul by the medium of another. But since the essence of the soul is compared to the powers both as a principle active and final,

development of the hylomorphic view of the mind-body relationship specific to the cogitative power is the proper response to Brentano's critique. The active intellect is simply the movement from what is potentially all things to the intellect becoming all things according to the mode of the knower where the intellective power is that whereby the proper object of the intellect moves from sensibility to intelligibility in substantial unity in the cognitive process of assimilation. Sensible forms are likenesses of an external object according to the same specific nature as the object but different in being.

The cogitative power, in the act of perception, prepares the *per se* sensible forms and incidental *per accidens* (incidental) intentions for abstraction from which the light of the intellect abstracts the universal *quiddity* or *whatness* of *per se* and *per accidens* sensibles and the quiddity is assimilated to the intellect where that which becomes known is in the knower according to the mode of the receiver.[225] In this way, the possible intellect *becomes all things* and the agent active intellect *makes all things* in the act of assimilation.

---

and as a receptive principle, either separately by itself, or together with the body; and since the agent and the end are more perfect, while the receptive principle, as such, is less perfect; it follows that those powers of the soul which precede the others, in the order of perfection and nature, are the principles of the others, after the manner of the end and active principle. For we see that the senses are for the sake of the intelligence, and not the other way about. The senses, moreover, are a certain imperfect participation of the intelligence; wherefore, according to their natural origin, they proceed from the intelligence as the imperfect from the perfect. But considered as receptive principles, the more perfect powers are principles with regard to the others; thus the soul, according as it has the sensitive power, is considered as the subject, and as something material with regard to the intelligence" (*ST* I, q. 77, a. 7, body; emphasis added). For discussions of the reflexive, synthetic, and discursive operations of common sense see Alfred Leo White, Jr., *The Experience of Individual Objects In Aquinas A Dissertation*, (Washington, D.C.: The Catholic University of America, 1997), 64-81; also Edmund J. Ryan, *The Role of the "sensus communis" in the Psychology of St. Thomas Aquinas* (Cartheena, Ohio: The Messenger Press, 1951); also Stephen John Laumakis, "The 'sensus communis' and the Unity of Perception according to Saint Thomas Aquinas" (Ph.D. diss., University of Notre Dame, 1991), 74.

225 Aquinas explains: "Now in order that the sight know whiteness, it is necessary for it to receive the likeness of whiteness according to its specific nature, although not according to the same manner of being because the form has a manner of being in the sense other from that which it has in the thing outside the soul: for if the form of yellowness were received into the eye, the eye would not be said to see whiteness. In like manner in order that the intellect understand a quiddity, it is necessary for it to receive its likeness according to the same specific nature, although there may possibly not be the same manner of being on either side: for the form which is in the intellect or sense is not the principle of knowledge according to its manner of being on both sides, but according to its common ratio with the external object" (*ST* III q. 92, a. 1, body; emphasis added).

The Process of Abstraction and the Unity of Intelligibility

The process of abstraction is one continuous act, an act by which the possible intellect through the agent intellect and the *vis cogitativa* continue down into the senses in the process of abstraction from phantasms. Although there is only one act, for analysis it is necessary to draw various distinctions in the intellective process of abstraction. There are various aspects to this act of moving from what is potentially intelligible to what is actually intelligible. First, there is the abstraction of sensible forms by the external and internal senses. Second, there is the preparation of the phantasms by the cogitative power for abstraction by the agent intellect. Third, there is the process of abstraction by the agent intellect of the intelligible species, and the process of returning to phantasms for indirect knowledge of the singular.

The proper object of the possible intellect is the universal quiddities of things, and the particular quiddity is the proper object of the *vis cogitativa*. Therefore the possible intellect can only know the particular quiddity indirectly through a process of reflection upon the phantasm. In reflecting upon phantasms in order to know the singular, the phantasm is the means for presenting the object and not the object of reflection itself. The proper object of the reflexive act is the object itself and "in this reflection it is not necessary to know the phantasm itself in its entity but rather the object represented therein."[226] It is necessary in the reflexive intellective act to know the object represented in the phantasm, and this object represented in the phantasm is the proper object of indirect knowledge of the singular.

The <u>first</u> aspect of moving from what is potentially intelligible to what is actually intelligible is the abstraction of sensible forms by the external and internal senses. In this first case, "the sense faculties, as cognoscitive powers, require a corporeal organ and therefore their proper objects are forms existing as such in corporeal matter. Since matter is the principle of individuation, every sense faculty knows

---

226 Edward Quinlisk Franz, *The Thomistic Doctrine on the Possible Intellect* (Washington D.C.: CUA Press, 1950), 169.

only particular, individual, material things as particular."[227] The proper object of the senses are forms existing individually in corporeal matter, but unlike the intellect the senses do not know these forms independent of the individual matter in which they exist.[228] Although the senses do not become the matter from which the external or internal senses abstract the sensible forms, yet the senses can still be changed by the sensible matter as touch becoming hot when touching a hot object, or in seeing something red, cone cells near the center of the retina are stimulated "where absorption changes the shape of the pigment" triggering various other physiological changes.[229]

With regard to the proper sensible object of the external senses and when the external senses are not defective, the external senses do not err in regard to their proper objects. In the case where the external senses are not defective, the eye does not err when it sees color, hearing does not err when it hears a sound, the sense of smell does not err that it smells a particular odor, and so forth. However, when the senses combine various sensations such as sight and hearing, as in the case of common sensibles, one may err even when the senses themselves are not defective. For example, hearing the sound of thunder and seeing

227 Ibid., 134.
228 Ibid., 134.
229 Aquinas discusses visual perception as "spiritual" but also allows for change within the bodily organ itself when acted upon. Likewise, contemporary physiology describes the process in terms of physiological changes associated with the absorption of *photons of particular wavelengths* that change the shape of pigments and gated sodium channels that open or close due to *changes in voltage* across a membrane, which in turn reduces neurotransmitter inhibitors *allowing for signals about the visual stimulus to be sent* to the brain. Although a physiological process is evident, there remains a movement from photons of particular wavelengths to electro-chemical changes that result in signals being sent to the brain. Of course, an external material object is not sent directly to the brain, it is the photons of particular wavelengths that act as visual stimulus upon the photoreceptors within the retina whereby images are sent as signals to the brain and this is the "spiritual" process to which Aquinas would likewise refer based upon contemporary physiology. The "spiritual" change is not one of similarity in being but rather a common ratio that exists between the knowing subject and the external object (*ST* III, q. 92, a. 1, body; emphasis added). In the case of the external senses, visual images are stimulants acting upon a bodily organ. Against the position of Galileo and Descartes that held secondary qualities are not properties of an object and against Russell and Berkeley who held that primary qualities are merely perceptual, science not only suggests that a cedar tree is composed of subatomic particles but also that the cedar tree is a solid object that smells of cedar and the smell of cedar is intrinsic to cedar. Therefore secondary qualities are properties of an object and primary qualities are not merely perceptual but do in fact have their basis in existence (cf. Cecie Starr, *Biology concepts and applications* (United States: Wadsworth Publishing Company, 1997), 568).

lightening may lead one to conclude two separate events, one causing thunder and another causing lightening. Or, seeing a small object in the distance and judging mistakenly that it is a man instead of a lion due to motion and distance.[230]

Likewise incidental sensibles, in judging that some white thing is sweet or that this individual who is white is the son of Diarus may be mistaken if one incorrectly has associated sweetness with a particular white object or associated mistakenly an image of the son of Diarus with a particular individual. The proper sensible of seeing something white is not mistaken, but the incidental association could be mistaken *although it does not have to be*. Likewise, one might mistakenly associate a given effect with a given cause. However, due to repeated associations, memory and experience in forming custom and habit decreases the opportunity that a mistaken sensible apprehension and association will take place.

In the second case, the cogitative power prepares the phantasm for abstraction. Franz describes the relationship between the intellect and abstraction from phantasms: "the proper object of the intellect are forms existing individually in corporeal matter, but the intellect knows these forms independent of the individual matter in which they exist."[231] Franz continues to explain, however, to know forms independent of their individuating conditions is to "abstract them from their individual matter in which they are represented by the phantasms or sense images. And, therefore, the human intellect understands material things by abstracting from phantasms."[232]

---

230 Such conditions of mistaken perception were attributed to the proper senses (or, secondary qualities) by the British Empiricists and Rationalists, and this of course led to the mistaken impression that sensation itself is predominately unreliable and therefore it became common practice to rely upon quantitatively established sensibles or primary qualities although these in fact are less reliable than proper sensibles (i.e., secondary qualities). The Empiricists and Rationalists of Early Modern philosophy simply failed to grasp the significance of incidental sensibles, failing to see that it is only by means of incidental sensibles that one can know the connection between an efficient cause and its effect since such events are singular and sensible.

231 Edward Quinlisk Franz, *The Thomistic Doctrine on the Possible Intellect* (Washington D.C.: CUA Press, 1950), 134; also see *ST* I, q. 85, a. 1.

232 Ibid., 134.

Prior to this process of abstraction from the phantasms, the cogitative power must prepare phantasms for abstraction. Because of the cogitative powers conjunction with the intellect and sensibility, it *reasons* upon the *particular* while the imaginative power combines images experienced through the senses in synthetic unity and these are stored in the memorative power.[233] At times this synthetic unity of the imagination may fall under the direction of reason where the *vis cogitativa* arranges and classifies the images in the imagination and the agent intellect acts by *directing* and *illuminating* the arrangement.[234] This arrangement is not that of the Kantian *a priori* construct, for neither the arrangement produced by the *vis cogitativa* nor the synthetic unity of images by the *phantasmata* are considered universal nor necessary since it falls under sensibility and the particular. However, the *synthetic unity of the appearances* or *empirical intuitions* under Kant is similar to the synthetic unity of the phantasmata although in the latter case the synthetic unity is entirely independent of the transcendental.[235]

---

233 Ibid., 122-123.

234 Ibid., 123.

235 For Kant experience includes both apprehending representations in sensibility and the activity of synthesis on the part of the imagination, and it is this activity of synthetic unity through which the manifold of sensible impressions are given unity and order. Experience is the product of this synthetic unity of the manifold of sensible impressions of the imagination. Hence, experience consists in intuitions or singular representations, but the imagination synthesizes or structures these intuitions allowing for generality. But it is here where Aquinas and Kant diverge in their understanding of synthesis within the imagination. For Kant the synthesis of the imagination proceeds according to schemata or rules, some of the rules are empirical while others are *a priori*. For Aquinas the imagination or phantasmata is only at the level of sensibility and not at the level of intelligibility and thus the synthesis that takes place in the imagination does not proceed according to schemata or rules, neither empirical nor *a priori*. Therefore for Aquinas the synthesis in the imagination is one of composition and division independent of the schemata. Further, this synthetic unity of the manifold of sensible forms never proceeds according to a given schema or rule, neither empirical nor *a priori*. For Kant the *a priori* rules or the *a priori* schemata are those by which intuitions represent causality, "representing an objective world of causally interacting substances standing in spatio-temporal relations to one another" (Hannah Ginsborg, "Thinking the Particular as Contained under the Universal" Draft Dept. of Philosophy, U. C. Berkeley, (November 2003), 4). For Kant there are also "rules or schemata corresponding to our empirical concepts, and it is in virtue of their accordance with these rules that our intuitions come to represent objects as having determinate empirical features, for example as having qualities like red or belonging to kinds like dog or house" (Ibid., 4). For Aquinas, there is no such rule at the level of the imagination even under the direction of understanding or intelligibility. Instead, Aquinas apprehends from existence the objective world of causally interacting substances standing in spatio-temporal relations to one another and conforming to the categories of being. Causality is an incidental sensible apprehended from sensibility rather than an *a priori* rule by which one represents an objective world of causally interacting substances standing in spatio-temporal relations to one another.

Independent of the transcendental in the sense of being independent of the rules for the imaginative synthesis of the manifold or simply independent of the rules or schemata some of which are empirical and others *a priori*.

In the case of the *vis cogitativa* acting upon images stored in memory, the *vis cogitativa* "collects and reasons on memorial images," those images stored in memory, and in doing so works upon experiences rather than images taken directly from the imagination.[236] These experiences stored

Klubertanz explains the process as understood by Aquinas as follows: "Once the three powers of imagination, memory, and the cognizant sense supervene upon the primary synthesis of the common sense, we can have the secondary, complete synthesis known as perception. For the cognizant sense unifies the combined data of the common sense, the imagination, and the memory, relating the object of sensation to its situation as a whole, as in the first instance, and to other experiences... Thus, from sense through memory (and so by repetition) there arises perception, the experimentum of St. Thomas (*In Meta*. IV, lect. 6; *In Post. Anal*. II, lect. 20, *In Meta*. I, lect. 1)....From the fully accomplished perception, which can even be called a 'quasi-universal' or 'implicitly universal knowledge, because in a way it includes a number of singular experiences, there arises the strictly universal and abstract knowledge which is the act of the intellect. The first concepts will naturally be quite general; later concepts will become more determinate, more specialized, in line with the law that intellectual knowledge proceeds from the general to the particular by intrinsic differentiation (cf. *ST* I, q. 85, a. 3)" (see George P. Klubertanz, "The Internal Senses in the Process of Cognition," *The Modern Schoolman* (January 1941), 30; emphasis added).

236 Aquinas explains the various cognitive and brain functions as follows: "For the reception of sensible forms, the "proper sense" and the "common sense" are appointed, and of their distinction we shall speak farther on (ad 1,2). But for the retention and preservation of these forms, the "phantasy" or "imagination" is appointed; which are the same, for phantasy or imagination is as it were a storehouse of forms received through the senses. Furthermore, for the apprehension of intentions which are not received through the senses, the "estimative" power is appointed: and for the preservation thereof, the "memorative" power, which is a storehouse of such-like intentions. A sign of which we have in the fact that the principle of memory in animals is found in some such intention, for instance, that something is harmful or otherwise. And the very formality of the past, which memory observes, is to be reckoned among these intentions. Now, we must observe that as to sensible forms there is no difference between man and other animals; for they are similarly immuted by the extrinsic sensible. But there is a difference as to the above intentions: for other animals perceive these intentions only by some natural instinct, while man perceives them by means of coalition of ideas. Therefore the power by which in other animals is called the natural estimative, in man is called the "cogitative," which by some sort of collation discovers these intentions. Wherefore it is also called the "particular reason," to which medical men assign a certain particular organ, namely, the middle part of the head: for it compares individual intentions, just as the intellectual reason compares universal intentions. As to the memorative power, man has not only memory, as other animals have in the sudden recollection of the past; but also "reminiscence" by syllogistically, as it were, seeking for a recollection of the past by the application of individual intentions. Avicenna, however, assigns between the estimative and the imaginative, a fifth power, which combines and divides imaginary forms: as when from the imaginary form of gold, and imaginary form of a mountain, we compose the one form of a golden mountain, which we have never seen. But this operation is not to be found in animals other than man, in whom the imaginative power suffices thereto. To man also does Averroës attribute this action in his book *De sensu et sensibilibus* (viii). So there is no need to assign more than four interior powers of the sensitive part---namely, the common sense, the imagination, and the estimative and memorative powers" (*ST* I, q. 78, a. 4; emphasis added).

in memory are synthesized by the *vis cogitativa* into the *experimentum* and this may be a "generic" image as John of St. Thomas suggests and the *experimentum* may be that which the *vis cogitativa* through frequent conversion and collation produces a phantasm more spiritual and more proximate to intelligibility as suggested by Cajetan.[237] In any case, from

Aquinas also explains that the intellect knows things that the senses cannot perceive such as incidental sensibles: "The intellect depends on the senses less than any power of the sensitive part. But the intellect knows nothing but what it receives from the senses; whence we read (Poster. i, 8), that 'those who lack one sense lack one kind of knowledge.'… Although the operation of the intellect has its origin in the senses: yet, in the thing apprehended through the senses, the intellect knows many things which the senses cannot perceive. In like manner does the estimative power, though in a less perfect manner…. the action of the cogitative power, … consists in comparing, adding and dividing, and the action of the reminiscence, … consists in the use of a kind of syllogism for the sake of inquiry,… The cogitative and memorative powers in man owe their excellence not to that which is proper to the sensitive part; but to a certain affinity and proximity to the universal reason, which, so to speak, overflows into them. Therefore they are not distinct powers, but the same, yet more perfect than in other animals" (*ST* I, q. 78, a. 4; emphasis added).

Further, Aquinas understands that mind, reason, understanding, and intellect to be one and the same power "Augustine says (Gen. ad lit. iii, 20) that "that in which man excels irrational animals is reason, or mind, or intelligence or whatever appropriate name we like to give it" (*ST* I, q. 79, a. 8). Therefore, reason, intellect and mind are one power. I answer that, Reason and intellect in man cannot be distinct powers." Aquinas also argues that "human reasoning, by way of inquiry and discovery, advances from certain things simply understood---namely, the first principles; and, again, by way of judgment returns by analysis to first principles, in the light of which it examines what it has found" (*ST* I, q. 79, a. 8; emphasis added). First principles such as the law of contradiction is apprehended inductively from experience and are of those things that are simply understood as one understands that a stone cannot both be and not be at the same time and in the same respect. First principles are inductively apprehended from experience and do not structure experience, but judgment by analysis returns to first principles by which judgment examines what it apprehends in the process of division and composition. 237 In the case of the internal senses, there is synthesis and perceptual integration and association that takes place between the *function* of memory, the *function* of common sense where the integration, coordination, and interpretation of various associated sensations takes place, the imaginative *function* that synthesizes images of individual sensible objects, and the cognitive *function* whatever the physiology might be that directs the various internal senses in preparation for abstraction of universals from sensible data, makes judgments at the sensory level, and manages incidental sensible associations. In the case of memory, contemporary physiology suggests that sensory input triggers memory and information encoded in sensory signals. Encoded signals from sensory input arrive at association centers in the cerebral cortex where they are packaged as chemical bonds and moved to various regions of the brain for storage. In contemporary physiology and biology, various theories are proposed on how the brain stores information. Based upon various experiments one finds, however, that long term memory persists and resists degradation whereas short-term memory is susceptible to Alzheimer's disease. *Habitus* and *experimentum* persist over time in sense memory as memory of the *past as past*. *Synderesis* understood as the habit of principles regulative of practical action can persist over time, and by analogy, long term memory allows for persistent intellectual memory and recollection of universals that go beyond time: "the possible intellect, in knowing the universal goes beyond time…. Retention and recall, belong *per se* to intellective memory as well as to sense… Such a memory is not, however, a faculty distinct from the possible intellect but simply a function of the latter" (Franz, 94, 108; *ST* I, q. 79, a. 6, ad. 2; *ST* I, q. 79, a. 1, 6c; *SCG* I, 56; *In De Anima* III, 1.8; *SCG* II, 74; *De Veritate* II, q. 10, a. 2). (cf. Cecie Starr, *Biology concepts and applications* (United States: Wadsworth Publishing Company, 1997), 560).

the fully accomplished perception that includes a number of singular experiences, arises through the process of abstraction the strictly universal and abstract knowledge which is the act of the intellect.

White indicates, "Aquinas, following Aristotle, identifies four stages in the development of knowledge. <u>First</u> comes the sensation of something; and from sensation comes the memory of that object which is the <u>second</u> stage. After one has retained the memory of several things that are the same in some way, the cogitative power can compare them to each other. This comparison results in the <u>third</u> kind of <u>cognitive act, which is called experience</u>. Through this cognitive act, one grasps (*accipit*) something that is common to the many individuals being compared in memory."[238] Hence, experience is directed toward something common to many individuals at different places and times, but it does not extend to all individuals. However, "after reasoning about many experiences of the same thing, one eventually arrives at the intellectual knowledge of first principles of art and science, which is the <u>fourth</u> stage of cognitive development."[239] One may recognize that a given medication commonly cures a certain ailment, and is thus common to many particular cases. White explains the process:

> The person who experiences an herb as having cured many people of the same malady does not simply pick out the same sort of herb from the many instances in his or her memory: instead, he or she recognizes both the herb and the fever as being common to many particular cases. Furthermore, one perceives an order between the two objects given in experience where one thing is affected by the other.... Experience does not perceive its complex object as a mere sequence of independent events [as Hume supposed]. Such would be the case if one simply noted that the application of a certain type of remedy is followed later by someone's becoming better. The experienced

---

238 Alfred Leo White, Jr. *The Experience of Individual Objects In Aquinas A Dissertation* (Washington, D.C.: The Catholic University of America, 1997), 200-202; emphasis added.

239 Ibid., 202.

person instead apprehends the herb itself as a source of healing [one apprehends that the source itself is the cause of its effect and it is this perceived relationship between individuals by the cognitive act that is the relationship or *connexion* between a given effect and its cause], or as possessing the tendency to bring about healing...the cogitative power, perceives relationships between individuals...through experiential cognition. In apprehending that the same herb has cured many individuals in the past, it does not consider only how this herb might be applied in the present situation for some particular future end; instead experiential cognition considers how many previous actions have led to the same result.[240]

White explains further that "experience brings about ease and correctness through habit, which in turn is caused by custom... Custom is a disposition of the memory to recall objects in a certain order; moreover, it shows that this tendency enables the cogitative and estimative powers to perceive relationships between different objects of perception."[241] Hume argued that a necessary *connexion* is an essential element in causal relations, but when one examines such relations one only finds contiguity and succession between a cause and its effect and never the *connexion* itself.

Contrary to Hume, custom does in fact allow one to perceive relationships and not simply a sequence of independent events. Humans acquire experience by comparing many singular intentions located in memory, and the disposition of custom is formed by a deliberate, discursive process of comparing many singular intentions and various relationships between singular intentions.[242] White argues following

---

240 Ibid., 203-205.
241 Ibid., 205-206; emphasis added.
242 Ibid., 207-208. Aquinas explains the process in some detail: "Now for complete sense knowledge, which would be adequate for an animal, five things are indispensable. First, that a sense power receive a species from sensible things, and this activity belongs to a proper sense (one of the external senses). Secondly, that there be a sense to discriminate among the sensible qualities perceived and to distinguish them from one another; and this action must be performed by a power in which all sensible perceptions terminate, and this power is called the unifying sense. Thirdly, that the species of sensible thing which have been received be retained. For an animal needs to know sensible things not only when they are

Aquinas: "Hence experience predisposes one to act correctly and easily by establishing custom in the memory."[243]

The cogitative power grasps that which is common to many similar individuals and the phantasm and memory represent a determinate nature to the intellect, but the determinate nature is only potentially intelligible. A determinate nature is a common nature existing according to material conditions and is discovered by experience and custom. This determinate nature, which is grasped by the cogitative power is not strictly speaking a universal since it is of the singular and sensible, but in another way "sense is in a certain way even of the universal."[244] For in apprehending the singular Callias, one also apprehends this human, and distinguishes this human from that human. Aquinas explains: "if it were such that sense only apprehended that which belongs to the particular, and in no way apprehended with this the universal nature in the particular, it would not be possible that from the apprehension of sense there should be caused in us the knowledge of the universal."[245]

---

present, but also after they are no longer present. And it is necessary that this activity be attributed to another power, because in corporeal things the principle which receives and that which retains are distinct; for that which is very receptive is sometimes poorly retentive. Now this power is called imagination or fantasy. Fourthly, a sense is required which might apprehend intentions that the other senses do not perceive, such as the harmful, the useful, and other notions of this sort. Now a human being arrives at a knowledge of these intentions by investigation and deliberation; but other animals possess this kind of knowledge by natural instinct, as, for example, a sheep naturally flees a wolf as being harmful. Hence in animals other than human beings a natural estimative power is directed toward this end, whereas in a human being there is a cogitative power, which compares these particular intentions; hence this power is called both the particular reason and the passive intellect. Fifthly, complete sense knowledge requires that things which were previously apprehended by the external senses and have been retained in the interior senses be once again summoned up for actual consideration. And this activity belongs to the power of recollection, which in animals other than human beings operates without investigation, but in human beings operates through inquiry and endeavor. Hence there is in human beings not only memory but also reminiscence. Now it was necessary that the power which is ordered to this end be distinct from the other powers, because the activity of the other sensitive powers involves a movement from things to the soul, whereas the activity of the power of recollection involves a movement from the soul toward things" (*Quastio de Anima* q. 13 Body; emphasis added; all citations from *Quastio de Anima* are from *Questions on the Soul*, trans. by James H. Robb (Milwaukee: Marquette University Press, 1984) unless otherwise noted).

243 Alfred Leo White, Jr. *The Experience of Individual Objects In Aquinas A Dissertation* (Washington, D.C.: The Catholic University of America, 1997), 208.

244 *In Post. Anal.* II, 20 (*Exposition on the Posterior Analytics of Aristotle*. Translated by Pierre H. Conway. Quebec: *La Libraire Philosophique* M. Doyon, 1956; also see Alfred Leo White, Jr. *The Experience of Individual Objects In Aquinas A Dissertation* (Washington, D.C.: The Catholic University of America, 1997), 210-211.

245 *In Post. Anal.* II, 20 (*Exposition on the Posterior Analytics of Aristotle*. Translated by Pierre H.

White explains that the object of sensory cognition includes the determinate nature, that is, that which is common to other individuals, and Aquinas explains that sensation must grasp what is common to individuals in order for the intellect to be able to know the universal.[246]

This "common nature" or the "universal nature in the particular" refers to "the nature itself, which may exist according to material conditions or may exist in the intellect as a universal."[247] In the process of abstraction, the intellect abstracts the intelligible species of a natural thing from individual sensible matter, but not from common sensible matter. For example, although one can abstract away these individual bones from the abstract universal concept man, one does not abstract by the intellect the common sensible matter of flesh and bones from what it means to be man.

Thus what is common to many such as flesh and bones are known both by the cogitative power as individual sensible matter (this flesh and these bones), and also by the intellect as common sensible matter simply as flesh and bones. Thus in sensibility, the individual sensible matter is apprehended and in the intellect both the universal and common sensible matter are abstracted from their individual sensible conditions allowing one to perceive the singular and know the universal. For example, bones are abstracted by the intellect leaving behind this or that set of bones, and man is abstracted by the intellect leaving behind the material conditions of this or that individual man. The cogitative power is "aware of something common to many not only when comparing

---

Conway. Quebec: *La Libraire Philosophique* M. Doyon, 1956). Similarly: "For it is clear that sensing is properly and per se of the singular, but yet there is somehow even a sensing of the universal. For sense knows Callias not only so far forth as he is Callias, but also as he is this man; and similarly Socrates, as he is this man. As a result of such an attainment pre-existing in the sense, the intellective soul can consider man in both. But if it were in the very nature of things that sense could apprehend only that which pertains to particularity, and along with this could in no wise apprehend the nature in the particular, it would not be possible for universal knowledge to be caused in us from sense-apprehension" (*In Post. Anal.* II, 20 from *Commentary on the Posterior Analytics of Aristotle*, trans. by F. R. Larcher (Albany, NY: Magi Books, 1970)).

246 Alfred Leo White, Jr. *The Experience of Individual Objects In Aquinas A Dissertation* (Washington, D.C.: The Catholic University of America, 1997), 211.

247 Ibid., 212. Also see *In De Anima* II, chpt. 12 (Leonine ed., 45.1:115.83-85).

many individuals that have been perceived in the past, but also when perceiving something that is present."[248]

Third, there is the process of abstraction by the agent intellect and degrees of abstraction of the intelligible species. As noted earlier, the human intellect understands material things by abstraction. The result of this abstraction process is the intelligible species, and this species is a likeness of the form in a thing and matter since both matter and form compose substantial being.[249] Franz explains the need for intelligible matter and form in the process of abstraction: "the intelligible species (*species intelligibiles*) is the likeness not of the form alone but also of the matter since both matter and form belong to the natural species in the real order of things. Therefore the intelligible species presents the definition of a thing or what a thing is. In other words, the intelligible species is an intellectual likeness of the essence or quiddity of a thing. Since the essence of every material thing must include matter, the intelligible species must be a likeness of matter in some way."[250] Franz indicates that the obvious problem is that abstraction is abstraction from matter and yet one asserts that essence is a likeness of matter.[251] However, the solution according to Aquinas as expressed by Franz is the following:

> Matter is twofold: common, e.g., flesh and bone, and individual (*signate*), e.g., this flesh and this bone. In the process of abstraction the intellect abstracts from individual sensible matter but not from common sensible matter so that in abstraction the species of man the intellect abstracts from this flesh and this bone (since these are not of the formality of the species but only of the individual) but does not abstract from flesh and bone. Therefore the intellect abstracts the *essence* of a thing. It is said to abstract the "form" because this is

---

248 Ibid., 213.
249 Edward Quinlisk Franz, *The Thomistic Doctrine on the Possible Intellect* (Washington D.C.: CUA Press, 1950), 135.
250 Ibid., 135.
251 Ibid., 135.

the principle of actuality in every essence, ie., the dynamic, active, determining principle. But since it must determine *something*, it must always be conceived and actually exist in relation to some matter for it to be a substantial form.[252]

Abstraction is of form and matter since form determines matter, and the intellect therefore abstracts from *signate* particular matter (i.e., this bone or this flesh) but not from common matter (i.e., bone and flesh).[253] Common matter itself is universal when conceived by the intellect. Franz also points to the fact that St. Thomas maintains that abstraction is demanded due to materiality and not due to individuality or singularity.[254] Franz continues:

> In abstracting from the individuating material conditions one still retains individual existence since essence must connote existence when speaking about natural things for both are contained in being (*ens*) … Thomistic philosophy… recognizes existence as the necessary correlative of essence insofar as it is the actualization of essence. Existence is related to essence as act to potency. Since potency and act must be in the same order, so must essence and existence. Therefore existence is not a predicamental accidental modification of essence but is the intrinsic dynamic principle of actuality within being (*ens*) which actualizes the essence.[255]

The <u>first degree of abstraction</u> carried out by the first operation of the intellect is the abstraction of "being (*ens*) as abstracted from individual sensible matter but not from common sensible matter (e.g., man is conceived as flesh and bone but not as this flesh and this bone—flesh and bone is common sensible matter; this flesh and

---

252 Ibid., 135.
253 ST I, q. 85, a. 1, ad. 2.
254 "It must be said that the singular is opposed to intelligibility not insofar as it is singular, but insofar as it is material because nothing is understood except immateriality" (ST I, q. 86, a. 1, ad. 3). See "Dicendum quod singulare non repugnat intelligenti inquantum est singulare, sed inquantum est materiale; quia nihil intelligitur nisi immaterialiter" (ST I, q. 86, a. 1, ad. 3; Franz, 137-136)
255 Edward Quinlisk Franz, The Thomistic Doctrine on the Possible Intellect (Washington D.C.: CUA Press, 1950), 136.

this bone is individual sensible matter)."[256] Comprehension (or the essence of a thing) and extension (or common matter) are abstracted from individual sensible matter. Thus being (*ens*), where *ens* is both existence and essence, is abstracted from individual but not common matter. The first degree of abstraction abstracts the natural species (i.e., *species naturalia* since this intelligible species is abstracted from natural things).[257]

The second degree of abstraction abstracts from both "individual and common matter and also individual intelligible matter but not from common intelligible matter (e.g., quantity, number, extension as abstracted from this material thing)."[258] Thus common sensibles or "primary qualities" are abstracted in this second degree of abstraction and the species formed is "designated mathematical species because they have to do with quantity, number, and extension."[259] The third degree of abstraction apprehends "being (*ens*) abstracted from all matter whether sensible (both individual and common) or intelligible (both individual and common) e.g., being *qua* being" and from this the transcendentals.[260] The intelligible species formed by this third degree of abstraction is called the metaphysical species because "they are a likeness of being in all its richness and purity insofar as it is devoid of all matter and its potentialities."[261]

Being is an analogous term and insofar as being is abstracted in the various degrees of abstraction, it is abstracted analogously and not univocally or equivocally.[262] Being in the first and third degrees of abstraction is real being (*ens reale*) since real existence is placed in its definition. The second degree of abstraction, however, is not real being but ideal being (*ens rationis*) because real existence is not placed in its definition but only imaginary existence. For example, mathematical

---

256 Ibid., 148.
257 Ibid., 148.
258 Ibid., 148.
259 Ibid., 148.
260 Ibid., 148.
261 Ibid., 148.
262 Ibid., 148.

entities form logical or ideal entities (*ens rationis*) and not species of actually existing things as they exist. For example, to abstract a house in the first degree of abstraction is to abstract from something that does in fact exist. However, <u>to abstract a mathematical circle is to abstract mathematical unity or quantity from existence, but mathematical abstraction does not abstract the idealized perfect mathematical circle itself from existence since the idealized circle is an abstract entity formed by the intellect.</u>

Therefore in the third degree of abstraction there is not abstraction of logical or ideal entities (*ens rationis*), but real being (*ens reale*) since this level of abstraction does include existence. Man does exist and is abstracted from this or that man. Likewise, God does exist and the notion of God's existence is abstracted from his effects. Similarly, being *qua* being can be abstracted from this or that being or being existing common to all things. Kant mistakenly confuses the category of logical or ideal being with real being proposing that God cannot be known from his effects or from the sensible conditions of human knowledge while maintaining that God is a moral postulate.

God's existence became associated with *ens rationis* rather than *ens reale* since for Kant God in no way could be an object of possible experience, and this became the fundamental flaw in transcendental philosophy in the Kantian metaphysics of theism. By imposing *a priori* categories upon existence without abstracting such categories directly from sensible things as Aristotle had down, Kant imposed imaginary or logical or mathematical entities (*ens rationis*) upon existence reversing the natural order of mathematical abstraction. In other words, if one were to impose upon an existing circle the notion of an idealized circle, the existing circle becomes the idealized or imaginary circle. However, if one abstracts the imaginary circle or quantity from an existing circle, the existing circle remains and one has abstracted the quantity of an existing circle rather than imposed an idealized quantity or an idealized circle upon existence. This was of course a legacy of Continental

Rationalism continuing to influence Kant's noetic even after his turn toward empiricism.

In Kant adopting the Cartesian mathematical model and reversing the order of intelligibility, Kant imposed logical entities (*ens rationis*) such as the Wolffian logical categories upon existing entities (*ens reale*). Consequently, one does not have a *common ratio with the external object*, instead one only has a manifold of appearances and no access to the *noumena* nor any apprehension of God *ens reale* where God's existence is abstracted from his effects and sensibility. Kant in drawing upon the mathematical method adopts the method of the Cartesian Rationalists in the Kantian transcendental method and thereby Kant detached further sensibility from intelligibility reversing the very order of knowability or knowledge acquisition.[263]

Aquinas explains, under the question *whether the judgment of the intellect is hindered through suspension of the sensitive powers* in *ST* I q84, a8, not only is the human intellect's proper object the nature of sensible things, but perfect judgment cannot be formed about the nature of something unless what *pertains to that things nature be known*.[264]

---

263 In Artificial Intelligence, knowledge acquisition is typically queried from an external user and it is this information that provides the basis for learning and knowledge content using heuristic models or neural nets for example. Likewise, when a child is learning a language, a parent points to this or that object and the child learns what each object is referred to by name rather than the child imposing upon existence such names or categorical relationships as it subjectively or logically imagines. Neither is this by custom in manner that Hume supposed since a child can learn the name 'father,' 'mother,' or 'puppy' by seeing such an object only once and that object being named. However, it may also be the case that many instances might also be required. If more than one association is required then it is required to fix such an association in memory and imagination where cognition can make proper associations. It seems likely that objects and relationships are abstracted from sensibility through experience either as intentional forms or incidental intentions where various relationships, resemblance, and contiguity occurs both external to the knowing subject and by means of perception and cognition.

264 Aquinas explains: "Our intellect's proper and proportionate object is the nature of a sensible thing. Now a perfect judgment concerning anything cannot be formed, unless all that pertains to that thing's nature be known; especially if that be ignored which is the term and end of judgment. Now the Philosopher says (*De Coel.* iii), that "as the end of a practical science is action, so the end of natural science is that which is perceived principally through the senses"; for the smith does not seek knowledge of a knife except for the purpose of action, in order that he may produce a certain individual knife; and in like manner the natural philosopher does not seek to know the nature of a stone and of a horse, save for the purpose of knowing the essential properties of those things which he perceives with his senses. Now it is clear that a smith cannot judge perfectly of a knife unless he knows the action of the knife: and in like manner the natural philosopher cannot judge perfectly of natural things, unless he knows sensible things. But in the present state of life whatever we understand, we know by comparison to natural

Without perceiving the *individual* as it is in itself independent of the Kantian transcendental, one cannot attain the existence nor the nature or essential properties of sensible things. What is the thing-in-itself except the essential properties and those properties common between objects as abstracted from those things which we perceive with our senses.

The reason why one cannot attain the existence nor the nature or essential properties of sensible things in the transcendental philosophy is that what is imposed upon the object by the transcendental philosophy is *ens rationis* or logical entities formed by the transcendental. The object itself is not apprehended or abstracted from *ens reale* therefore all that is known are *a priori* logical entities since it is not the individual as a thing-in-itself that is attained, rather all that is known is the *synthetic unity of the manifold of appearances* according to rules for the imaginative synthesis of the manifold and not the *things-in-themselves* in the case of the Kantian transcendental. This was essentially Kant's solution that Aquinas would have rejected as being inconsistent with experience and a reversal of the very order of knowability.

Since the sensible object as it is in itself is unknown in the transcendental philosophy, comprehension understood as the essence, nature, or the essential properties of the individual is never attained. Following the *via moderna*, all that is attained is extension or similarity in synthetic judgments between appearances as determined by the Kantian *a priori* rules of the transcendental or the empirical schemata. What can be known of an individual is preconditioned by epistemic synthetic *a priori* rules in judgment. For example, the universality of man as a rational animal *always or for the most part* (for the most part since one may be born with a disability or experience brain damage) is necessarily the case for Aquinas by means of abstracting the essence or nature of man in comprehension through the phantasms of sense knowledge of the form or nature of each individual man rather than being preconditioned by epistemic synthetic *a priori* rules that condition such necessity in judgment.

---

sensible things. Consequently it is not possible for our intellect to form a perfect judgment, while the senses are suspended, through which sensible things are known to us" (*ST* I q. 84, a. 8, body).

A judgment is thus formed 'man is a rational animal' where the verb copula joins the predicate. Thus the essential nature or properties of what it means to be man is necessary and universal and thus *a priori*, but this *a priori* nature is abstracted from the universal and necessary essential nature in this or that man. It is this once abstracted that becomes the abstract universal subject man. However, for Kant such universality and necessity is preconditioned by the *a priori* rules of space and time and the categories as projected upon individuals in the *synthetic unity of the manifold of appearances* where preconditioned similarity and *a priori* rules determine what is known of the individual. Thus the judgment 'man is a rational animal' is preconditioned for Kant by similar conditions imposed by the knowing subject upon similar external appearances between individuals in the *synthetic unity of the manifold of appearances* and not from the very nature or form existing in this or that man.

For Kant it is by extension or from the early modern bundle theory rather than comprehension that the essence or forms of sensibility are apprehended as constructed by the *a priori* transcendental method. Thus intelligibility rather than sensibility becomes the determinate condition for the knowing subject similar to a Platonist epistemology and developed for similar reasons, the unreliability of sense perception, a contingent view of existence, and a need for universality and necessity. However, for Aquinas, universality and necessity can be abstracted from sensibility being the objects of intelligibility, and the end term, or purpose of judgment about the nature of something is to know the *essential properties* of those things passively received by the senses, and *one cannot know these essential properties of things-in-themselves apart from the senses and independent of the transcendental method and the preconditioned manifold of appearances*:

> Our intellect's proper and proportionate object is the nature of a sensible thing. Now a perfect judgment concerning anything cannot be formed, unless all that pertains to that

thing's nature be known; especially if that be ignored which is the term and end of judgment. ... the natural philosopher does not seek to know the nature of a stone and of a horse, save for the purpose of knowing *the essential properties of those things which he perceives with his senses*. ... in like manner the natural philosopher *cannot judge perfectly of natural things, unless he knows sensible things*. But in the present state of life whatever we understand, we know by comparison to natural sensible things. ... (emphasis added).[265]

Kant in reversing the order between intelligibility and sensibility eliminated the possibility of the intellect to abstract from the sensible objects themselves the essential properties of those things which the senses perceive, and therefore science cannot judge of natural things because one cannot know the essential properties. Rather than saving science from the failed psychology of the British Empiricists and the Continental Rationalists and Hume's critique, Kant destroyed scientific knowledge by eliminating the epistemological basis for knowing the essential qualities and definition of natural things from which science using the experimental method discovers. This was not a failure attributed to science since applied science continued down the path of using the experimental method, but it was a failure in philosophical epistemology and another move toward Cartesian subjectivism.

In contrast, cognition terminates in the sense, in imagination, or in intellective power alone in three degrees of abstraction. For Aquinas "in the first degree of abstraction, the senses present the properties and accidents of a thing so as to express sufficiently the nature of a thing which the intellect abstracts and which the intellect judges to be true by terminating in the data so presented by the senses. Thus the cognition of the natural sciences terminates in the senses which have presented the data through observation and experimentation."[266] The second

---

265 *ST* I, q. 84, a. 8 Body.
266 Edward Quinlisk Franz, *The Thomistic Doctrine on the Possible Intellect* (Washington D.C.: CUA Press, 1950), 150.

degree abstracts from common sensibles the mathematical entities such as quantity, extension, etc. and the intellect judges not based upon the senses perceiving sensible matter but upon data abstracted from common sensibles as logical entities even though in existence these logical mathematical entities are "concretized in sensible matter."[267] Thus in the second degree of abstraction, cognition terminates in imagination or the phantasm. It is in the imagination that a physical line is apprehended as a quantitative abstract line, or a circular object can be understood in terms of a perfect circle. "Therefore in mathematical judgment the intellect must terminate in the imagination which presents an ideal line, point, and circle which do not exist as such in reality."[268]

The third degree of abstraction, exceeds all that fall under the scope of the senses and that which is abstracted from sensible matter in phantasms, which are "used as principles from the data of which the intellect arrives at metaphysical concepts either through causality (in which the cause is deduced from the effects), through excess (in which perfections are deduced as eminently perfect), or through remotion (in which this perfection is deduced by removing all imperfections)."[269] Metaphysical causal relationships such as the judgment that a cause is deduced from its effects exceed those principles from which the intellect arrives at such concepts (i.e., the senses and phantasms in apprehending efficient causal events and incidental relationships), but those principles derived from incidental sensibles for example provide the basis for metaphysical concepts. As the incidental sensible of an efficient cause and effect relationship is abstracted, the causal relationship provides the basis for a metaphysical concept that a cause is deduced from its effect as will be discussed in more detail in the following chapter.

There is also a distinction between judicial and simple abstraction in Aquinas. Judicial abstraction occurs according to the manner of

---

267 Ibid., 150.
268 Ibid., 151.
269 Ibid., 150.

composition and division, and simple abstraction is the process of abstraction where one thing is apprehended without something that pertains to it. For example, in judicial abstraction one may judge that Socrates is composed of rationality and animality, and his essential nature is not that of irrationality. In the case, of simple abstraction, one apprehends man without the individuating conditions of "this flesh" and "these bones" but not independent of common matter such as flesh and bones.

In the case of simple apprehension, there is no possibility of falsity because one abstracts the universal by eliminating any individuating material conditions. For example, flesh without this flesh and bone without these particular bones. There is no judgment, assessment, or evaluation formed at the level of concept formation. However, in the case of composition and division, one can err in judging that one thing belongs to another. For example, one can judge that true gold is false, or that Socrates has green eyes rather than brown in returning to the sensibles conditions of human knowledge, or one can judge that the essential nature of man is irrationality due to participation in animality.[270] E. Q. Franz explains this process of abstraction as follows:

---

270 Aquinas writes: "Abstraction may occur in two ways: First, by way of composition and division; thus we may understand that one thing does not exist in some other, or that it is separate therefrom. Secondly, by way of simple and absolute consideration; thus we understand one thing without considering the other. Thus for the intellect to abstract one from another things which are not really abstract from one another, does, in the first mode of abstraction, imply falsehood. But, in the second mode of abstraction, for the intellect to abstract things which are not really abstract from one another, does not involve falsehood, as clearly appears in the case of the senses. For if we understood or said that color is not in a colored body, or that it is separate from it, there would be error in this opinion or assertion. But if we consider color and its properties, without reference to the apple which is colored; or if we express in word what we thus understand, there is no error in such an opinion or assertion, because an apple is not essential to color, and therefore color can be understood independently of the apple. Likewise, the things which belong to the species of a material thing, such as a stone, or a man, or a horse, can be thought of apart from the individualizing principles which do not belong to the notion of the species. This is what we mean by abstracting the universal from the particular, or the intelligible species from the phantasm; that is, by considering the nature of the species apart from its individual qualities represented by the phantasms. If, therefore, the intellect is said to be false when it understands a thing otherwise than as it is, that is so, if the word "otherwise" refers to the thing understood; for the intellect is false when it understands a thing otherwise than as it is; and so the intellect would be false if it abstracted the species of a stone from its matter in such a way as to regard the species as not existing in matter, as Plato held. But it is not so, if the word "otherwise" be taken as referring to the one who understands. For it is quite true that the mode of understanding, in one who understands, is not the same as the mode of a thing in existing: since the thing understood is immaterially in the one

The agent intellect shines its 'light' upon the phantasm in which the object is presented as potentially intelligible and thereby makes the object actually intelligible by presenting to and within the possible intellect the intelligible species which is the object on the intellectual plane. The possible intellect when informed by this intelligible species, generates within itself the *intentio, verbum,* or concept which is the spiritual intentional likeness of the object. This process is known as abstraction by the possible intellect. The intelligible species produced in the possible intellect by the combined action of the object and the agent intellect is the *principium* for the generation of the *verbum* in the possible intellect as the *terminus.* The process of abstraction is completed with the formation of the *verbum.*

The agent intellect serves the possible intellect by producing the intelligible species which serves the formation of the *verbum.* The intelligible species is called the impressed species and is necessitated by the fact that the object is only potentially intelligible before the formation of this species. It is also necessitated by the fact the possible intellect is a *receptive* faculty which requires information in order to become activated. The *verbum* is called the expressed species and is necessitated by the fact that the possible intellect is an *operative* faculty whose nature it is to generate its knowledge according to the information of the object through the impressed species... the possible intellect knows the object through the impressed and expressed species and knows the species themselves through reflexion. The species is *that through which (id quo)* the object is known and not *that which (id quod)* is known (*ST* I, q85, a2).[271]

---

who understands, according to the mode of the intellect, and not materially, according to the mode of a material thing" (*ST* I q. 85, a. 1, ad1).

271 Edward Quinlisk Franz, *The Thomistic Doctrine on the Possible Intellect* (Washington D.C.: CUA Press, 1950), 141-142.

This describes how the universal is abstracted from the singular in St. Thomas, but Franz also explains the process by which knowledge of singulars or particulars is possible: "In reflecting upon the phantasm for the knowledge of particulars the possible intellect cannot by itself complete the reflection. The action of the *vis cogitativa* must enter into play so that there may be a complete 'return' to the particular as particular (*De Anima* a20, ad 1)."[272] The intellect mixes itself with singulars and continues within the reflection of the *vis cogitativa* as an intellective operation.[273] Franz explains further:

> In the knowledge of singulars, the *vis cogitativa* knows this or that singular not insofar as it apprehends the matter as matter in these individuals, but insofar as the individuating notes have some relationship to the form as individuated. These individuating notes can be understood only insofar as the *vis cogitativa* 'partakes' of the intellect… the agent intellect must function along with and within the *vis cogitativa* in order for this highest of sense faculties to apprehend the form through which it understands individuating notes as such. It is only in the 'light' of the agent intellect and in the light of 'being' that these notes are intelligible to the possible intellect…When the possible intellect reflects back upon the intelligible species to know the singular, it reflects not upon the species as such, but upon the *object* presented there through the abstraction of the agent intellect.[274]

Franz explains that the origin of the intelligible species is the agent intellect as the principal efficient cause, the *vis cogitativa* as the instrumental efficient cause, and the phantasm as the formal objective cause representing the individual.[275] The agent intellect through the *vis cogitativa* brings to light the material quiddity represented in the

---

272 Ibid., 167.
273 *De Veritate* II, q. 10, a. 5.
274 Edward Quinlisk Franz, *The Thomistic Doctrine on the Possible Intellect* (Washington D.C.: CUA Press, 1950), 167.
275 Ibid., 168.

phantasm, and through the *vis cogitativa* the possible intellect mixes with singulars insofar as it apprehends the particular through the action of this intellective operation. Franz explains, "the *vis cogitativa* completes the process through the action of the agent intellect which must be strictly correlated not only with the intelligible species it has produced in the possible intellect, but also with the action of the *vis cogitativa* in apprehending the particular material quiddity [which is represented in the phantasm]."[276]

It should also be noted that this process of returning to the phantasm to obtain knowledge of the singular is one continuous act or operation of the intellect bridging between intelligibility and sensibility.[277] In judgment likewise the intellective operation reflects back down to the senses, but in addition to this reflexive act, in judging that 'Socrates is a man,' the intellect apprehends the relation between the supposite 'Socrates' and the species 'man' and assents to their agreement.[278] The judgment reflects upon the singular when it assents to the agreement that the singular 'Socrates' is a 'man' in existence.[279]

In answering the questions raised by Brentano, the historical investigation of the internal senses demonstrate a continued development and reformulation informed by the cognitive sciences in the Aristotelian tradition. Aquinas's development of the hylomorphic view of the mind-body relationship, the Dionysian principle of hierarchy, and the reformulation of the notion of *continuatio* suggest a composite unity where there are no gaps between sensibility and intelligibility in response to Brentano's critique.[280] Although not directly tied to any particular physiology, the cognitive psychology

---

276 Ibid., 168.
277 Ibid., 169.
278 Ibid., 169.
279 Ibid., 169.
280 An indication that such a unity exists is that the intellect itself can become weary because the powers that employ the organs of the body are subject to fatigue, and it is not possible for the intellect to have thought without employing such organs: "We become weary in mind after long or concentrated thought, because powers that employ organs of the body are subject to fatigue, and in this life it is not possible to give the mind to thought without employing those organs" (SCG IIIa, 62).

of Aquinas lends itself to contemporary cognitive psychology and physiology. Intentionality, the process of abstraction, the relation between universals and particulars, sensible and intelligible forms and intentions, and the cognitive psychology of Aquinas have been discussed in response to Brentano's critique given earlier, and this in preparation for discussing how synthetic *a priori* knowledge of efficient causality is possible.

Chapter 5

SYNTHETIC *A PRIORI* KNOWLEDGE OF EFFICIENT CAUSALITY

Universal and necessary principles of contingent things consequent of form in the subject are known by the intellect by abstracting necessary conditions consequent of form or incidental intention from particular matter. For example, what consequently follows from Socrates sitting is the necessity that Socrates sits if Socrates is in fact sitting. Likewise, what follows necessarily from the form of man individuated in Socrates is that Socrates is necessarily a rational animal. The senses know the singular Socrates as contingent, but if Socrates exists and is a man then the intellect knows the universality and necessity that Socrates is a rational animal consequent of the form man. Similarly in regard to common sensibles, it follows that if Socrates runs, Socrates necessarily moves, and the relation of running to motion is a necessary common sensible, for Socrates must necessarily move if he runs. This common sensible necessary relationship is abstracted from the contingent conditions of sensibility in the process of abstraction.

Since this necessity of motion follows from the common sensible of Socrates being in motion, and since the intention of Socrates being in motion is abstracted through the common sensibles from the contingent particular of Socrates running, the necessity of motion is abstracted as a universal. Therefore, the intellect knows universal and necessary principles of contingent things as universal and necessary. Due to the synthetic unity that occurs both at the level of perception and at the level of judgment, synthetic knowledge of universality and necessity is possible. In the case of efficient causality taken as an incidental sensible, the intellect is directed to the *intentio* from which a universal and necessary causal relationship or *connexion* is formed. The cogitative power i.e. cognition in forming an incidental intention from the singular prepares the intention for abstraction by the intellect in the formation of the intelligible species or the quidditative concept of a universal and necessary causal relationship in the first operation of the intellect.[281]

---

281 Specifically in contemporary cognitive physiology the cogitative power as an operation of the intellect can be associated with the cerebrum responsible for sense perception, thinking, consciousness, memory, and discursive judgment or learning. It should also be noted that contemporary cognitive psychology maintains at least four stages of cognitive development following Piaget's theory. The first phase is the sensorimotor phase (from birth to 2yrs) that gives the ability to coordinate sensory input with motor actions. At the end of this stage a child can use mental images or phantasms to represent objects (e.g., a mental image of a favorite toy) and the child is able to recognize that objects continue to exist even when they are no longer available to the external senses. Due to the activity of recognizing that objects continue to exist when removed, the use of mental images to represent absent objects develop. Using mental images to represent an absent object is the primitive stage of symbolic thought being acquired by adding and removing objects from a child's sensory input formed from memory, custom, and imagination and thus *object permanence* develops. The second is the preoperational stage (age 2 to 7) is marked by an *inability* to judge that physical quantities remain constant in spite of changes in their shape or appearance. For example, the same quantity of water when poured into a taller and thinner beaker is grasped by the child to be more in the taller thinner beaker. The child judges that the thinner beaker has more water than the other beaker due to the higher water line. At this stage children are unable to 1. focus on several aspects of the problem at once, 2. unable to undo something, and 3. and they associate living qualities with inanimate matter.

The concrete operational stage (age 7 to 11) is where a child can only perform operations on tangible images and actual events. The cogitative power or the cerebrum continues to develop where singulars are grasped and judged. For example, they can sort daisies and carnations but when asked if there are more flowers than carnations they respond that there are more carnations than flowers because they are unable to abstract a higher level of abstraction where flower is the genus of carnations and daisies. However, they do at this stage grasp the categories of being and various common sensibles such as mass, number, volume, area, and length or quantity and quality which they were unable to grasp in the preoperational stage. They also grasp logical relations such as Bill is younger than Sally, and Sally is younger than Jenny. Thus the cognitive operation at the concrete stage is marked by internal transformations, manipulations, reorganizations of cognitive structures apprehended directly from existence and is specific to concrete operations. This suggests that at this stage a clear relationship between the cogitative power and that of the intellect has developed but the emphasis is upon the singular concrete tangible images and actual events rather than abstract concepts. The formal operational stage is the final stage (begins around 11 years of age) where a child applies their operations to abstract concepts in addition to concrete concepts moving from the second stage of abstraction to the third stage of abstraction. Abstract concepts such

Quidditative concepts contain the abstracted efficient cause, the efficient effect, and the incidental intentional *connexion* that unites the cause with its effect. Just as form and common matter are abstracted, the incidental intention and common matter are abstracted in the case of incidental sensibles. Thus an efficient cause is abstracted from any material conditions such as this ball, this bat, this broken window, and what is abstracted is the efficient causal events and the relationship between a given cause and its effect as an abstract incidental association or relationship.

The second operation of the intellect combines a given cause with its effect based upon the incidental intentional causal relationship or *connexion* judged to exist by the cogitative power and abstracted by the intellect. Thus the cogitative power judges at the level of singularity and contingency that this ball and this bat broke this window. Likewise, in abstraction at the level of universality and necessity the intellect judges the universal and necessary proposition 'ball and bat broke window'. This is one intellectual operation where the cognitive faculty (i.e., the cerebellum in contemporary physiology) determines that this ball and this bat broke this window therefore the bat necessarily broke the window and the intellect itself judges that any bat and ball can universally and necessarily brake a window given the right conditions.

Hence the intellect concludes in a universal and necessary judgment that it is always or for the most part (unless hindered) the case that a window can be broken by a ball and bat given the right set of circumstances, and that this window was broken by this ball and this bat as the intellect returns to the cogitative judgment and phantasm or to the external senses. Similarly, in model propositions where the antecedent condition and its consequent are abstracted from

as love, justice, free will can be apprehended abstracted from actual events and tangible images or phantasms. In adolescents one becomes more systematic at this stage, inductive trial-error approaches to problem solving develop as well as deductive approaches of hypothetical problem solving occurs using logic to reason out likely consequences and solutions. Thus at this stage thought processes in the formal operational stage are abstract, systematic, reflective, and logical. Cf. Wayne Weiten, *Psychology* 4th ed. (Mexico: International Thomson Publishing Inc., 2000), 321-324.

experience, one deduces the necessary consequent from the antecedent conditional. For example, one apprehends that this ice table is made of ice and the intellect abstracts the necessary conditional 'if the ice table is made of ice, it is necessarily made of ice' and cannot be otherwise so long as it is made of ice *in rebus*. At the same time the necessary consequent of being made of ice is likewise abstracted from experience since the ice table is necessarily made of ice and cannot be otherwise so long as it is made of ice *in rebus*.

To the degree that a given effect can be universally and necessarily reunited with its cause, one has a universal and necessary causal relationship. Once a universal and necessary causal relationship is determined to exist by an act of the intellect in forming a judgment where the cause and its effect are reunited in a universal and necessary causal relationship, that universal and necessary causal relationship judged to exist by the intellect can be reapplied in the process of reflection through the cognitive faculties (cogitative power) to any singular efficient causal relationship or modal necessity in existence to determine its necessity and universality. If the particular or singular efficient causal event occurs always or for the most part, the intellect judges that a necessary relationship exists between a given efficient cause and its effect due to contiguity and an incidental causal *connexion*.

In so doing, synthetic *a priori* knowledge of efficient causality is possible since a necessary connection is found to exist between a given cause and its effect as apprehended by the synthesizing role of cognitive perception and the intellect in a single cognitive operation. Further, by judging from many instances of efficient causal events, one can form the causal maxim that every effect has a cause. For instance, by apprehending that it is the case that every window that is broken has a cause for it to be broken and this is the case with x number of events, one can form from this the general causal maxim that every effect has a cause. The following sections will explain this process in some detail.

Apprehension and Judgment

The question to be addressed in this section is how through the process of abstraction can one apprehend universal and necessary causality from efficient causality apprehended from sensibility? In the process of abstraction, there are three operations of the intellect for Aquinas. The three degrees of abstraction discussed in the previous chapter occur in virtue of these three operations of the intellect.

The <u>first operation</u> is simple apprehension where an object is grasped without affirming or denying anything of the object. It is here where concept formation occurs. Wallace indicates that simple apprehension seeks distinct and clear knowledge of objective concepts by grasping the comprehensive notes and extensive parts of these objective concepts.[282]

The <u>second operation</u> is judgment where something is affirmed or denied of something else. It is here where subject and predicate affirmation or denial occurs, or where simple and complex sentences, judgments, propositions, and modal propositions are formed, and where truth as conformity between what exists and what is known is determined through affirmation or denial. It is at this level that categorical and modal propositions are formed and therefore modal claims of necessity/impossibility and contingency/possibility obtain.

The <u>third operation</u> of the intellect is discursive reason where one proceeds from previous knowledge to new knowledge in virtue of syllogistic reasoning. It is at this level that syllogistic necessity and scientific knowledge occurs. The following will discuss the first and second operations of the intellect in an effort to explain the process of abstraction. Hence simple concept formation in the first operation of the intellect will be discussed and judgment, propositions, identity, and truth and falsity in the second operation of the intellect will be explained in some detail.

In the first operation of the intellect, the essence and common matter are apprehended through the process of abstraction, and are

282 William A. Wallace, O.P., *The Elements of Philosophy* (New York: ALBA House, 1977), 17.

associated or dissociated with one another in the second operation or act of the intellect to form propositions and where a uniformity of being or a *common ratio with the external object* exists where identity conditions between a given subject and predicate take place between intelligibility and sensibility independent of the transcendental method. Independent of the transcendental philosophy in the sense of being independent of the rules for the imaginative synthesis of the manifold or simply independent of the rules or schemata some of which are empirical and others are *a priori*.

Wallace explains, "to know the thing as it is in reality, a single whole, one and concretely existing, a mental operation is required that reintegrates the intelligible aspects of the thing and signifies it as existing. This requires a comparison, the establishment of a relation, which is the unity of its terms... By judging and forming a proposition the intellect restores natures to subjects and accidents to substances, thus re-establishing the condition in which things exist."[283]   In the medieval logic of the *via antiqua*, there was a metaphysical component that allowed for a uniformity of being at the level of the verb copula where things united in judgment were to be united in such a way as they exist in reality. As seen in Figure 2 below, in Thomistic metaphysics there is a double intentionality of being which is the *real being toward the knower, and its complement, the intentionality of the knower back toward its source in real being.*[284]

The first operation of the intellect or simple apprehension is that by which the intellect forms simple concepts of things by understanding the whatness or quiddity or the essence (*essentia*) of each thing (*res*) as the first act of the intellect grasps what a thing is, i.e., its essence (*essentia*) or quiddity, without affirming or denying anything of it. Henle indicates that there are two distinct questions to ask about real things: "What is it?" and "Is it?"[285] The first question is often answered by describing certain characteristics such as it is blue, circular, etc. as an

---

283 Ibid., 17.
284 W. Norris Clarke, S.J., "Reflections on John Deely's Four Ages of Understanding," *International Philosophical Quarterly* Vol 43, No. 4, Issue No. 172, (December 2003), 536-537.
285 R. J. Henle, S. J. *Theory of Knowledge* (Chicago: Loyola University Press, 1983), 50-55

act of apprehension.[286] The second is an act of judgment determining that something is.[287]

Again drawing from Henle the Latin term *quid* means *what*, and the aspects that express a given thing are called quiddities.[288] The concept in which one understands quiddities are called quidditative concepts or the intelligible species. The reality of an existing thing expressed in the concept is a quiddity and includes both the essence or nature of a thing as well as common matter (not this flesh and these bones, but flesh and bones) allowing the quiddity to apprehend the individuality abstracted from material conditions.[289] The act of knowledge is the judgment or the second operation of the intellect, and judgment is the act by which some kind of existence is understood. For example, "woman is rational," or "man is," or "man is an animal that walks on two legs, laughs, and is rational."

Norris Clark indicates that the phenomenology of Brentano is a rediscovery of scholastic intentionality where "consciousness of its nature is directed toward another."[290] However, a complementary aspect of medieval intentionality lost to phenomenology is that movement of the knower back to the object from which knowledge of the object came *would not even be possible apart from the object communicating its own act of existence to the knowing subject in ontological intentionality* where the object makes itself known by actively projecting an intentional similitude, or sign of itself, into the knower through cognitive intentionality.[291] As illustrated in Fig. 2 below, it is this sign or similitude, which the knower can then interpret as a formal sign pointing back toward its source, thus affirming that the content of a given formal concept is the object which is known in virtue of an external object's being (*ens*) communicating its own act of existence (*esse*) and its essence (*essentia*)

---

286 Ibid.
287 Ibid.
288 Ibid.
289 Ibid.
290 W. Norris Clarke, S.J., "Reflections on John Deely's Four Ages of Understanding," International Philosophical Quarterly Vol 43, No. 4, Issue No. 172, (December 2003), 536-537.
291 Ibid., 536-537.

to the knower. Hence, the cognitive intentionality or the intelligible species, functions as both the *formal concept* as the means 'by which' (*id quo*) the thing known is understood and also the *objective concept* as 'that which' (*id quod*) is understood as containing the quiddity or the whatness of a thing as illustrated in Fig. 2.

Aristotle's "semantic triangle" given in his *On Interpretation* allowed the *via antiqua* to formulate a conception of signification where words immediately signify the concepts of the mind in intentional being (*esse intentionale*), and it is this mediation of the concepts whereby these concepts signify things as illustrated in Fig. 2. Thus words only have a nominal existence and do not reflect the mode of being that something has in existence except indirectly since it is the intellect through complex concepts that understand the mode of being that something has in existence. If the words arranged in sentences or statements correctly reflect a true proposition in the intellect, words will represent the way things are in existence. However, if the words or statements reflect a false proposition formed by judgment, these words, statements, or sentences will likewise be false. At the level of judgment truth or falsity obtain and statements, sentences, or propositions reflect this truth or falsity in judgment.

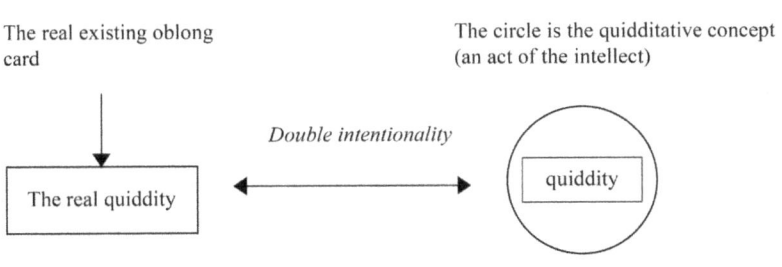

The real existing oblong card

The circle is the quidditative concept (an act of the intellect)

Double intentionality

The real quiddity

quiddity

Figure 2 *Conceptio*

The universal oblong shape is the quiddity contained in the quidditative concept. The intellect is thus assimilated to the oblong card and it is the oblong card that is directly known by the intellect and only by reflection is the quidditative concept itself understood by the intellect.

Aquinas explains: "words do not signify things directly according to the mode of being which they have in reality, but indirectly according to the mode in which we understand them; for concepts are the likeness of things, and words the likenesses of concepts, as is stated in Book I of the *Periherminias*" (*In VII Metaph.*, lect. 1, no. 1253). Hence our words signify whatever we can think, whether it exists external to a knowing subject or not. Even though beings of reason may not have any existence independent of the knower, beings of reason (*entia rationis*) have their own intentional existence in the knower. Thus logical and mathematical abstract complex concepts can reflect truth or falsity as given in intellectual judgment.

When speaking of universal expressions, sentences, statements, they signify the conceptions of the intellect and only through these conceptions do vocal sounds, sentence, statements signify things. For example, the name 'man' is an abstract universal concept signifying in abstraction human nature and does not immediately signify a singular man. For words to signify the particular or singular, the intellect through the cogitative power must reflect upon the singular using memory including custom and experience, phantasms, and/or the common and external sense. Therefore, words cannot immediately signify the singular but must indirectly signify the singular through concepts and when necessary by returning to the sensible conditions of human knowledge. Aquinas explains the relationship between universal concepts and language:

> Therefore 'passions of the soul' must be understood here as conceptions of the intellect, and names, verbs, and speech signify these conceptions of the intellect immediately according to the teaching of Aristotle. They cannot immediately signify things, as is clear from the mode of signifying, for the name 'man' signifies human nature in abstraction from singulars; hence it is impossible that it immediately signify a singular man. The Platonists for this reason held that it signified the

separated idea of man. But because in Aristotle's teaching man in the abstract does not really subsist, but is only in the mind, it was necessary for Aristotle to say that vocal sounds signify the conceptions of the intellect immediately and things by means of them.[292]

Not only do words signify concepts, Aquinas also indicates that there is a three fold diversity between things signified by names. First, beings outside the knower, such as a stone or a man. Second, something only in the knower, such as dreams or the imagination of a chimera. And third, things that have a foundation external to the knower, where the formal account (*ratio*) is determined or perfected by an operation of the knower, such is the case with universals. Aquinas explains, "for humanity is something in reality, but there it is not universal, for there is no some humanity outside the soul common to many, but as it is conceived by the intellect, by the intellect's operation a further concept [*intentio*] is adjoined to it, on account of which it is called a species" (*In Sent.* I, d19, q5, a1).[293] Universals are abstractions existing in this or that individual as individuated, but not as existing in this or that singular as universal. Human nature as a universal is an abstraction from this or that human nature existing as individuated in Socrates or Plato as

---

292 *Aristotle on Interpretation: Commentary by St. Thomas and Cajetan,* trans. by J. T. Oesterle (Milwaukee: Marquette University Press, 1962), 25.

293 Aquinas explains in detail: "It remains, then, that human nature happens to have the character of a species only through the being it has in the intellect. Human nature has being in the intellect abstracted from all individuating factors, and thus it has a uniform character with regard to all individual men outside the soul, being equally the likeness of all and leading to a knowledge of all insofar as they are men. Because it has this relation to all individual men, the intellect discovers the notion of species and attributes it to the nature. This is why the Commentator says that it is the intellect that causes universality in things. Avicenna makes the same point.

Although this nature apprehended by the intellect has the character of a universal from its relation to things outside the soul, because it is one likeness of them all, nevertheless as it has being in this or that intellect it is a particular apprehended likeness. The Commentator was clearly in error then; he wanted to conclude that the intellect is one in all men from the universality of the apprehended form. In fact, the universality of this form is not due to the being it has in the intellect but to its relation to things as their likeness. In the same way, if there were a material statue representing many men, the image or likeness of the statue would have its own individual being as it existed in this determinate matter, but it would have the nature of something common as the general representation of many men." *On Being and Essence* Ch 3 Para 7 [*Aquinas on Being and Essence,* trans. by A. A. Maurer (Toronto: PIMS, 2nd ed., 1968), 48]

a rational animal.[294] The definition of the essence gives the intelligible structure of the individual and the idea or quidditative concept is the intelligible image of the very nature of the object understood.

In order to form a judgment one must have something about which to judge, and this something may exist (a chair) or this something may be thought in some possible or imaginary sense. *Judgment* itself is the act of *assertion* of identity or nonidentity expressed in propositional terms by the verb copula. The function of the copula is to unite or disunite the two terms of the proposition in judgment. Judgment is that by which two concepts are brought together and joined by a copula, and this by virtue of the comparison between simple concepts (either concrete or abstract concepts). The judgment in the act of uniting the many and signifying the union or reintegration of the real natures *represented* by the concepts, identify the predicate with the subject as understood in existence.

This union is not merely uniting two terms, but it is called *concretion* where two come to exist together as one and exercise one act of being forming one identity, and as one being, there is numerical identity of reason when the predicate is joined to the subject and thereby becoming one and the same being in reason. In existence there remains a distinct act of existence between the knower and thing, but in virtue of the formality of the notes and the essential nature that make up the

---

294 Aquinas explains that humanity apprehended without consideration of individuality is considered the universal and brought under the intellect, as a likeness of a specific nature; however, taken as the specific nature in the individual it is the nature of that individual and not a universal: "the thing actually understood there is a double implication---the thing which is understood, and the fact that it is understood. In like manner the words "abstract universal" imply two things, the nature of a thing and its abstraction or universality. Therefore the nature itself to which it occurs to be understood, abstracted or considered as universal is only in individuals; but that it is understood, abstracted or considered as universal is in the intellect. We see something similar to this in the senses. For the sight sees the color of the apple apart from its smell. If therefore it be asked where is the color which is seen apart from the smell, it is quite clear that the color which is seen is only in the apple: but that it be perceived apart from the smell, this is owing to the sight, forasmuch as the faculty of sight receives the likeness of color and not of smell. In like manner humanity understood is only in this or that man; but that humanity be apprehended without conditions of individuality, that is, that it be abstracted and consequently considered as universal, occurs to humanity inasmuch as it is brought under the consideration of the intellect, in which there is a likeness of the specific nature, but not of the principles of individuality (*ST* I, q. 85, a. 1, reply 2; emphasis added).

definition (comprehension or connotation), there is identity: the same formal notes or attributes are in the concrete particular as individuated, and the knower in its universality as expressed by its connotation.[295]

This identity means unity in substance both in existence, as that which is a distinct and separate thing as a real being (*ens reale*) in existence, and in intelligibility expressed materially by a proposition.[296] It is this unity in nature and definition, which is identity. The universal nature and the specific notes of its definition applied to many concrete particulars become one intentionally and this one intentional being expressed by the proposition is the same as the nature and attributes existing in one composite substantial being in existence when judged truly.

Therefore it is not merely a relation nor is it merely extensional in the sense of logical connectives such as "and", "or", and "not". There

---

295 Similarly Mark Delp explains: "Comprehension or connotation signifies the number or variety of notes that make up an overall definition, the extension signifies an indefinite number of individuals, or kinds of individuals. For example, the universal nature in the intellect signified by the verbal name 'tree' embraces in its comprehension the notes 'woody perennial plant,' 'having a single main stem with no branches on the lower part,' and 'crowned with foliage', whereas in its extension it signifies a certain nature realized in an indefinite number of kinds of trees such as oak, alder, pine, birch, and spruce. Therefore, the comprehension or connotation *rather than the extension* of a concept is synonymous with its definition, which in turn is made up of the intelligible characteristics that make it what it is, i.e., some essence or nature" apprehended intentionally but still existing in things. On the other hand, the extension or denotation of a concept signifies that nature as it is manifest in existing things as it exists in any number of individuated singulars. Thus the universal nature in the intellect connoted by the verbal name 'tree' is individually applied to and exists in concrete particulars as denoted in this oak and in this pine, etc." (Delp, Logic Reader, 2005).

296 Aquinas explains that the combination and separation in the intellect is itself traced to a thing's disposition as its cause, and in predication there is identity in combination and division: "you are not white because we think truly that you are white; but conversely we think you are white because you are white. Hence it has been shown that the way in which a thing is disposed is the cause of truth both in thought and in speech. He adds this in order to clarify what he said above, namely, that in things truth and falsity consist in being combined and being separated. For the truth and falsity found in speech and in thought must be traced to a thing's disposition as their cause. Now when the intellect makes a combination, it receives two concepts, one of which is related to the other as a form; hence it takes one as being present in the other, because predicates are taken formally. Therefore, if such an operation of the intellect should be traced to a thing as its cause, then in composite substances the combination of matter and form, or also the combination of subject and accident, must serve as the foundation and cause of the truth in the combination which the intellect makes in itself and expresses in words. For example, when I say, "Socrates is a man," the truth of this enunciation is caused by combining the form humanity with the individual matter by means of which Socrates is this man; and when I say, "Man is white," the cause of the truth of this enunciation is the combining of whiteness with the subject. It is similar in other cases. And the same thing is evident in the case of separation" (*In Meta.* II, 9.11.1897-1898; emphasis added).

is a metaphysical uniformity of being, i.e., a *common ratio with the external object* (*ST* III, q92, a1, body), that takes place in the act of judgment, where judgment puts back accidents into their substances and the given nature back into the intentional subject from which it was abstracted as a metaphysical act of being.[297] The word "conception" is used in order to signify that in the word of our intellect is found the likeness of the thing understood.

The essence is either the thing itself or a given cause known from its effects: "the essence of a thing is either the thing itself, or is related to it in some way as cause: since a thing derives its species from its essence" (*SCG* 1.21). Although of course there is no numerical identity of nature or being between the external object and the quiddity, there is a uniformity of nature or being or a *common ratio between the knowing subject and the external object*. The ten categories are structures apprehended from existence rather being Kantian transcendental conceptual categories existing as logical entities independent of the way things are in themselves. Instead, the Aristotelian categories are a structure by which the intellect apprehends existence from existence,

---

297 Aquinas indicates that judgment belongs to metaphysics, mathematical abstraction from sensibility belongs to mathematics, and abstraction of the universal from the particular to all the sciences but particularly to the physical sciences: "We conclude that there are three kinds of distinction in the operation of the intellect. There is one through the operation of the intellect joining and dividing which is properly called separation; and this belongs to divine science or metaphysics. There is another through the operation by which the quiddities of things are conceived which is the abstraction of form from sensible matter; and this belongs to mathematics. And there is a third through the same operation which is the abstraction of a universal from a particular; and this belongs to physics and to all the sciences in general, because science disregards accidental features and treats of necessary matters. And because certain men (for example, the Pythagoreans and the Platonists) did not understand the difference between the last two kinds of distinction and the first, they fell into error, asserting that the objects of mathematics and universals exist separate from sensible things" (*The Division and Methods of the Sciences* q. 5, a. 3 Body; emphasis added). In Descartes and the early moderns adopting the Pythagorean and Platonist mathematical model in epistemology, they failed to understand properly mathematical abstraction and the abstraction of universals from particulars, and thus Kant came to assert that like the Pythagorean mathematical model, that the *a priori* transcendental can attain universality and necessity without direct abstraction from sensible things. However, objects of mathematics and universals do not exist entirely separate from sensible things. Although such objects can be known by the intellect as abstracted from sensibility, yet quantity, extension, and number are abstracted from material things when first apprehended while idealized mathematical entities are formed by the intellect from such abstractions. Similarly, the Kantian transcendental categories are abstracted from sensible things, although the categories are known as abstracted from sensibility. For the proper method of mathematical abstraction, see *The Division and Methods of the Sciences* q. 5, a. 3 Reply 1.

consistent with and according to existence.[298] The so called pre-philosophical or pre-Kantian categories express things as they are in themselves independent of the Kantian transcendental philosophy. The synthetic *a priori* is simply universality and necessity found in intelligibility as abstracted from sensibility.

Thus one finds that if we are to know the real thing (*res*) as it is, or as an imaginary creature might be in the imagination, as one intentional subject and one being, one must see it reintegrated, with accidents in the substance and various accidents joined in the one composite being and this is expressed materially in propositional form.[299] Schmidt explains, "in the proposition the subject will represent the thing which is known, and the predicate will represent what we know about it.

---

298 Aquinas explains, "We must realize (with the Philosopher) that the term 'a being' in itself has two meanings. Taken one way it is divided by the ten categories; taken in the other way it signifies the truth of propositions. The difference between the two is that in the second sense anything can be called a being if an affirmative proposition can be formed about it, even though it is nothing positive in reality. In this way privations and negations are called beings, for we say that affirmation is opposed to negation, and that blindness is in the eye. But in the first way nothing can be called a being unless it is something positive in reality. In the first sense, then, blindness and the like are not beings.

The term 'an essence' is not derived from this second meaning of 'a being', for in this sense some things are called beings that do not have an essence, as is clear in the case of privations. Rather, 'an essence' is derived from 'a being' in the first meaning of the term. As the Commentator says, a being in the first sense of the term is that which signifies the essence of a thing. And because, as we have said, 'a being' in this sense is divided by the ten categories, essence must mean something common to all the natures through which different beings are placed in different genera and species, as for example humanity is the essence of man, and so with regard to other things" (emphasis added). *On Being and Essence* Ch 1 Para 3 [*Aquinas on Being and Essence*, trans. by A. A. Maurer (Toronto: PIMS, 2nd ed., 1968), 30].

299 Aquinas explains: "Humanity, however, is that which is signified by the definition of man, as the essence of anything whatever is that which is signified by its definition. But the definition of man signifies not form alone but also matter, since matter must be comprised in the definition of material things. Hence we shall do better to say, with others, that both soul and body are included in the notion of humanity [Avicenna, Metaph., V, 5 (89va)], although otherwise than in the definition of man. The notion of humanity embraces only the essential principles of man, prescinding from all other factors. For, since humanity is understood to be that whereby man is man, whatever cannot truly be said to constitute man as man, is evidently cut off from the notion of humanity. But when we speak of man, who has humanity, the fact that he has humanity does not exclude the possession of other attributes, for instance, whiteness, and the like. The term "man" signifies man's essential principles, but not to the exclusion of other factors, even though these other factors are not actually, but only potentially, contained in the notion of man. Hence "man" signifies as a whole, per modum totius, whereas "humanity" signifies as a part, per modum partis, and is not predicated of man. In Socrates, then, or in Plato, this determinate matter and this particular form are included. Just as the notion of man implies composition of matter and form, so if Socrates were to be defined, the notion of him would imply that he is composed of this flesh and these bones and this soul" (emphasis added). *Compendium of Theology* Prt 1 Tr 1 Ch 154 [*Compendium of Theology*, trans. by Cyril Vollert (St. Louis: Herder, 1947), 164].

Thus the subject will stand for the supposit, and the predicate will stand for some quiddity or nature which belongs to that supposit and which in some way we find in it."[300] Aquinas explains these two modes of abstraction:

> For if we understood or said that color is not in a colored body, or that it is separate from it, there would be error in this opinion or assertion. But if we consider color and its properties, without reference to the apple which is colored; or if we express in word what we thus understand, there is no error in such an opinion or assertion, because an apple is not essential to color, and therefore color can be understood independently of the apple. Likewise, the things which belong to the species of a material thing, such as a stone, or a man, or a horse, can be thought of apart from the individualizing principles which do not belong to the notion of the species. This is what we mean by abstracting the universal from the particular, or the intelligible species from the phantasm; that is, by considering the nature of the species apart from its individual qualities represented by the phantasms (emphasis added).[301]

And yet if one does consider abstraction in reference to sensible things, the intelligible species is understood secondarily upon reflection to be that by which the intellect understands.[302] The intelligible species

---

300 Robert W. Schmidt, *The Domain of Logic According to Saint Thomas Aquinas* (Netherlands: Martinus Nijhoff, 1966), 226.

301 *ST* I, q. 85, a. 1, Rp. 1.

302 Merely knowing impressions or intuitions independent of knowing the objects themselves necessarily leads to the early modern conception of the clarity of ideas and the association of ideas or names rather than the association of the objects themselves. This naturally leads to subjectivism.

Aquinas explains: "Some have asserted that our intellectual faculties know only the impression made on them; as, for example, that sense is cognizant only of the impression made on its own organ. According to this theory, the intellect understands only its own impression, namely, the intelligible species which it has received, so that this species is what is understood. This is, however, manifestly false for two reasons. First, because the things we understand are the objects of science; therefore if what we understand is merely the intelligible species in the soul, it would follow that every science would not be concerned with objects outside the soul, but only with the intelligible species within the soul; thus, according to the teaching of the Platonists all science is about ideas, which they held to be actually understood [*Summa Theologiae* Q84, A1]. Secondly, it is untrue [that the intellectual faculties know only the impression made on them], because it would lead to the opinion of the ancients who maintained that "whatever seems, is true" [Aristotle, *Metaph.* iii. 5], and *that consequently contradictories are true simultaneously....*

is the likeness of an external object and upon reflection one realizes this to be the case, but the intelligible species is a likeness of the external object itself and it is this external object rather than a likeness of the external object that the intellect knows properly and primarily. Aquinas explains:

> The intelligible species is that which is understood secondarily; but that which is primarily understood is the object, of which the species is the likeness. This also appears from the opinion of the ancient philosophers, who said that "like is known by like." For they said that the soul knows the earth outside itself, by the earth within itself; and so of the rest. If, therefore, we take the species of the earth instead of the earth, according to Aristotle (*De Anima* iii, 8), who says "that a stone is not in the soul, but only the likeness of the stone"; it follows that the soul knows external things by means of its intelligible species. The thing understood is in the intellect by its own likeness; and it is in this sense that we say that the thing actually understood is the intellect in act, because the likeness of the thing understood is the form of the intellect, as the likeness of a sensible thing is the form of the sense in act. Hence it does not follow that the intelligible species abstracted is what is actually understood; but rather that it is the likeness thereof.[303]

and so every judgment will be true: for instance, if taste perceived only its own impression, when anyone with a healthy taste perceives that honey is sweet, he would judge truly; and if anyone with a corrupt taste perceives that honey is bitter, this would be equally true; for each would judge according to the impression on his taste. Thus every opinion would be equally true; in fact, every sort of apprehension" (*ST* I, q. 85, a. 2 body; emphasis added).

The modern subjectivism or skepticism that Kant himself attempted to avoid by introducing the transcendental philosophy is answered appropriately by the *via antiqua*. If one's subjective impressions are the basis for judging what is true then contradictory impressions that appear to be true are in fact true, which is false. Therefore modern subjective impressions or opinions cannot be the basis for judging what is true or false. Instead Aquinas explains that the intelligible species or impressions as likenesses of physically existing objects are understood only secondarily upon reflection, but what is primarily understood and that which provides an objective basis for truth claims are the objects themselves of which the intelligible species is simply a likeness. Therefore it is not the association of ideas or the clarity of ideas as ideas, but the association of the objects themselves that provide the basis for truth claims and not merely logical satisfaction.

303 *ST* I, q. 85, a. 2 body.

For Aquinas the intelligible species is the likeness of the thing understood, and the thing actually understood is the form of the external object and this becomes assimilated as the form of the intellect allowing for identity whereby the thing understood is the form of the intellect. Therefore what is understood in the primary sense is the thing itself since like is known by like just as the intellect knows the earth outside itself, by the earth within itself. This allows judgment to say that something is when it is, and it is not when it is not according to the ontological mode of the law of contradiction.

The quidditative concept serves as the intelligible species having intentional existence and taken as an abstract universal by which the real quiddities are known. One can bring together diverse quiddities into a whole to form an oblong green smooth light weight card, but still the list of quiddities are independent of existence until one asserts by means of the verb copula that the object along with all of its quiddities are real or existing. One grasps the existence of a subject along with all of its quiddities in judgments where identity or uniformity of being takes place according to composition and division and the law of contradiction, and one expresses this existence and this uniformity or identity in propositions or sentences. The quiddity that exists in the intellect has intentional existence, and the real quiddities existing in things has physical or ontological existence in things. Aquinas explains the kind of identity by which the intellect predicates one thing to another:

> For the <u>proper object of the human intellect is the</u> <u>quiddity of a material thing</u>, which comes under the action of the senses and the imagination. Now in a material thing there is a twofold composition. First, there is the composition of form with matter; and to this corresponds that composition of the intellect whereby the universal whole is predicated of its part: for the genus is derived from common matter, while the difference that completes the species is derived from the form, and the particular from individual matter. The second

comparison is of accident with subject: and to this real composition corresponds that composition of the intellect, whereby accident is predicated of subject, as when we say "the man is white." Nevertheless composition of the intellect differs from composition of things; for in the latter the things are diverse, whereas <u>composition of the intellect is a sign of the identity of the components</u>. For the above composition of the intellect does not imply that "man" and "whiteness" are identical, but the assertion, "the man is white," means that "the man is something having whiteness": and the subject, which is a man, is identified with a subject having whiteness. It is the same with the composition of form and matter: for animal signifies that which has a sensitive nature; rational, that which has an intellectual nature; man, that which has both; and Socrates that which has all these things together with individual matter; and <u>according to this kind of identity our intellect predicates the composition of one thing with another</u> (emphasis added).[304]

Identity, composition, or uniformity of being takes place as the intellect composes, for example, the composition of a subject with its accidents, or composing form with matter. The singulars are known indirectly while the universals are known directly, and thus the particular or singular Socrates can be predicated of the universal man by returning to the phantasm to apprehend the singular Socrates. Socrates is thus composed of the sensitive nature of animal, the intellectual nature of man, and individual matter. Sentences, statements, and language in general signify propositional judgments that return to sense knowledge for knowledge of the singular. Thus the predication of singulars of universals takes place by returning to sense knowledge or phantasms:

> Now what is abstracted from individual matter is the universal. Hence our intellect knows directly the universal only. <u>But indirectly, and as it were by a kind of reflection, it can</u>

---

304 *ST* I q. 85, a. 5 Rp 3.

know the singular, because, as we have said above (ST I, q85, q7), even after abstracting the intelligible species, the intellect, in order to understand, needs to turn to the phantasms in which it understands the species, as is said *De Anima* iii, 7. Therefore it understands the universal directly through the intelligible species, and indirectly the singular represented by the phantasm. And thus it forms the proposition 'Socrates is a man'" (emphasis added).[305]

Thus the singular or particular is known by returning to the phantasm where the singular is predicated of the universal concept man. Again, Aquinas makes clear that the intellect can do what the senses can, but in a more eminent fashion: "The higher power can do what the lower power can, but in a more eminent way. Wherefore what the sense knows materially and concretely, which is to know the singular directly, the intellect knows immaterially and in the abstract, which is to know the universal."[306]

In Lesson 4 of Book 6 of the *Commentary on Aristotle's Metaphysics* Aquinas discusses the role of the intellect in truth and falsity and explains that being which is true and non-being which is false depends upon combination and separation since simple concepts do not signify truth or falsity, but instead complex concepts, where one thing is predicated of another, have truth and falsity through affirmation and negation.[307] Combination or affirmation signifies that a predicate belongs to a given subject, and separation or negation signifies that a predicate does not belong to a subject.[308] Words are signs of concepts and as such, sentences or statements signify complex concepts that have truth or falsity through affirmation or negation. Therefore that which is true (being) and that which is false (non-being) depend upon combination and separation such that a predicate and subject can be related to an affirmation and a negation in two ways.

305 *ST* I, q. 86, a. 1.
306 ST I, q. 86, a. 1 Rp. 4.
307 In Meta. VI, 4.1223.
308 In Meta. VI, 4.1223.

First this composition is *connected in reality*, as man is connected to animal, and thus man is affirmed to be a species of animal in reality. Or, in the second way, they are *unconnected in reality*, "as man and ass" since man is not an *ass* in reality. Aquinas develops truth and falsity based upon contradictions of connected and unconnected terms at the level of judgment or propositional logic associated with the way things are in reality. For example, "man is animal" and "man is not an animal" is a contradiction but with connected terms.

Likewise, "man is an ass" and "man is not an ass" is a contradiction but with unconnected terms, "man" and "ass." Truth exists when there is an affirmation in the case of connected terms or when satisfaction obtains within both the proposition itself and in conformity to the way things are. Likewise, truth exists when there is negation in the case of unconnected terms or when satisfaction does not obtain within both the proposition and in conformity to a given state of affairs. For example, "man is an animal" and "man is not an ass" are both true since in the first case there is conformity and thus affirmation of connected terms and the latter case one has negation of unconnected terms. Falsity exists when the reverse is the case, when negation of connected terms exist or when affirmation of unconnected terms exist. For example, "man is not an animal" and "man is an ass" are both false because there is negation of connected terms in the former case, and because there is affirmation of unconnected terms in the latter case.[309]

Aquinas then explains that the truth and falsity is secondarily in the combination and separation of *words*, but primarily and properly truth and falsity is in the combination and separation of *concepts*, which the intellect makes. The intellect understands things combined or separated according as they are said to be together insofar as they form one thing. The intellect in combining and separating, understand both the subject and predicate as one thing, i.e., according as one thing is constituted from them. Similarly, the intellect understands the parts by understanding the whole. The intellect understands the whole

309 *In Meta.* VI, 4.1224-1226.

house together insofar as one thing is constituted from many parts. The intellect understands the whole in synthetic unity rather than a succession of parts, insofar as the synthetic unity is constituted from the parts. Similarly, the intellect understands a predicate and subject together insofar as one judgment is constituted from the subject and predicate as either combined or separated.[310]

Aquinas then explains the two operations of the intellect. The first operation is called *the understanding of indivisibles* or simple apprehension as discussed above, and is the operation whereby "the intellect forms simple concepts of things by understanding the whatness of each one of them. The other operation is that by which the intellect combines and separates."[311] St. Thomas explains that although truth and falsity pertains primarily and properly to the intellect and only secondarily to the combination and separation of words, truth and falsity do not pertain to that operation whereby simple concepts are formed nor do they pertain to the whatness of things.[312]

When the mind apprehends the whatness or essence of a thing, since there is no composition or division, there is no possibility of error at the level of simple apprehension or resulting from the first operation of the intellect where simple concepts are formed and the essence is abstracted by the intellect. Aquinas comments that Aristotle through this process of elimination indicates that it follows that truth and falsity is not in things, nor in "the mind when it apprehends simple concepts and the whatness of things, and therefore they pertain primarily and principally to the combination and separation which the mind makes, and only secondarily to that of words, which signify the mind's operation."[313]

---

310 *In Meta.* VI, 4.1227-1229.
311 *In Meta.* VI, 4.1232.
312 "Now while truth and falsity are in the mind [and only secondarily in the combination and separation of words [*In Meta.* VI, 4.1227], they do not pertain to that operation by which the mind forms simple concepts and the whatness of things. This is what he means when he says 'with regard to simple concepts and the whatness of things there is neither truth nor falsity in the mind'" (*In Meta.* VI, 4.1233).
313 *In Meta.* VI, 4.1233.

Knowing attains completion as a result of the likeness of the thing known coming to exist in the knowing subject and to the degree that the knowing subject possesses a correct likeness of the thing known that knowledge is true, and to the degree that the knowledge of the knowing subject falls short of that likeness that knowledge is said to be false.[314] Aquinas explains:

> Intellect has within itself a likeness of the things known according as it forms concepts of incomplex things, it does not for that reason make a judgment about this likeness. This occurs only when it combines or separates. <u>For when the intellect forms a concept of mortal rational animal, it has within itself a likeness of man; but it does not for that reason know that it has a likeness, since it does not judge that 'Man is a mortal rational animal.'</u> There is truth and falsity, then, only in this second operation of the intellect, according to which it not only possesses a likenesses of the thing known but also reflects on this likeness by knowing it and by making a judgment about it. Hence it is evidence from this that truth is not found in things but only in the mind, and that it depends upon combination and separation (emphasis added).[315]

In simple apprehension, where simple concepts are formed, the intellect forms a likeness of a man consisting of the essence and common matter of man united in a simple concept. However, the intellect does not affirm that it has this likeness until the essential elements or notes contained in the definition or essence of what a man is, is joined to man in the intellectual operation of reflection and in making a judgment about what man is. In the act of judgment, these are composed or separated by the verb copula signifying the act of existence. In this act of composition and division in judgment, the intellect determines if the likeness apprehended is in fact consistent with the way things are. In general terms, something must correctly conform to the thing

---

314 *In Meta.* VI, 4.1234.
315 *In Meta.* VI, 4.1236.

to which it is assigned for it to be considered true, and if it fails to conform then it is false.[316] Aquinas indicates this by stating:

> A false thing, as is said at the end of Book V, means one that does not exist in any way (for example, the commensurability of a diagonal) or one that exists but is naturally disposed to appear otherwise than it is. Similarly a definition is said to be false either because it is not the definition of any existing thing or because it is assigned to something other than that of which it is the definition... a thing is said to be true when it has the proper form which is shown to be present in it; and a definition is said to be true when it really fits the thing to which it is assigned.[317]

More specifically, falsity is either something in the intellect that does not exist in any way, or that exists but is naturally disposed to appear otherwise than it is as in the case of false gold. Likewise, a thing is said to be true if the form in the intellect is shown to be present in the thing, and a definition is true if it conforms to the thing to which it is assigned either in the intellect or in existence.

Aquinas draws again this connection by explaining: "combination and separation, on which truth and falsity depend, are found in the mind and not in things; and that if any combination is also found in things, such combination produces a unity which the intellect understands as one by a simple concept. But that combination or separation by which the intellect combines or separates its concepts is found only in the intellect and not in things. For it consists in a certain comparison of two concepts, whether these two are identical or distinct in reality."[318] In the case where a combination is found in things, the combination produces a unity that the intellect in the first operation of the intellect

---

316 This is similar to a truth value that obtains due to satisfaction; however, satisfaction occurs not at the level of names as in nominalism, but at the level of what is signified by those names as it exists external to a knowing subject. So a mathematical model is appropriate for logical second intentions but not for first intentions where names signify concepts and concepts signify things existing external to a knowing subject.

317 In Meta. VI, 4.1237.

318 In Meta. VI, 4.1241.

understands in a simple concept. In the case of the second operation of the intellect, judgment compares two concepts to determine if they are identical or distinct in fact. The next section applies apprehension of universality and necessity to knowledge of efficient causality and in judging that a given cause is connected to a given effect.

Knowledge of Synthetic a priori Efficient Causality

The previous section discussed the process of abstraction and judgment demonstrating that synthesis of simple universal and necessary quiddities are possible in the second operation of the intellect. One must further ask, how is knowledge of synthetic *a priori* efficient causality possible? How does the first and second operations of the intellect form universal concepts of causality and necessity? For Aquinas custom is a disposition of memory that allows objects to be recalled in a certain order and relationship, and "this enables the cogitative power to perceive relationships between different objects of perception."[319] Aquinas indicates according to White "that humans acquire experience by comparing many singular intentions in the memory."[320] Custom, however, does not compare many singular intentions and in this way experience differs from custom "inasmuch as a human acquires experience as a result of deliberately comparing individuals with each other."[321]

The intellect can draw relationships from the singular that are commonly the case and therefore universal. Thus one can experience that a particular medication has the same effect and based upon this common behavior the intellect can abstract the simple concept of a causal relation that is universal and necessary unless hindered. From this common behavior one can apprehend or abstract a universal and necessary causal relationship abstracted from incidental sensible conditions independent of the Kantian transcendental while avoiding Hume's critique. Independent of the transcendental philosophy in the

---

319 Alfred Leo White, Jr. *The Experience of Individual Objects In Aquinas A Dissertation*. Washington, D.C.: The Catholic University of America, 1997, 206.

320 Ibid., 206.

321 Ibid., 207.

sense of being independent of the rules for the imaginative synthesis of the manifold or simply independent of the rules or schemata some of which are empirical and others are *a priori*.

In the case of incidental sensibles such as causal relationships, the agent intellect is directed to the common *intentio* rather than the common *ratio* of external objects. The cogitative power prepares the incidental intention for abstraction with the aid of the intellective power, and the intellect abstracts the *intentio* to form the quiddity of the intelligible species. The quiddity is the necessary efficient causal *intentio* or *connexion* apprehended in the first operation of the intellect. The intelligible species contains the abstracted efficient cause, the efficient effect, and the incidental intention that unites the cause with its effect. The second operation of the intellect combines a given cause with its effect based upon the incidental intentional causal relationship and *to the degree that a given effect can be universally and necessarily reunited with its cause, one has a universal and necessary causal relationship.*

Although there are various senses of necessity, if taken in the absolute sense, a necessary causal event is a cause intimately and closely connected with its effect.[322] Aquinas commenting on Aristotle explains that the primary notion of necessity derives from the fact that something cannot be otherwise.[323] Therefore, a universal and necessary causal relationship within a given context cannot be otherwise than it is and is universally applicable given a particular context or set of conditions. Given the same set of conditions, the ball thrown through a breakable window will always and everywhere break the window.

322 *In Meta.* V, 6.833.

323 Aquinas explains: "Here he reduces all of the senses in which things are necessary to one...Hence that cause without which a thing cannot live or exist or possess a good or avoid an evil is said to be necessary; the supposition being that the primary notion of the necessary derives from the fact that something cannot be otherwise. Then he shows that the necessary in matters of demonstration is taken from this last sense, and this applied both to principles and to conclusions. For demonstration is said to be about necessary things, and to proceed from necessary things. It is said to be about necessary things because what is demonstrated in the strict sense cannot be otherwise...in demonstrations the premises are the causes of the conclusion, for demonstrations in the strict sense are productive of science and this is had only by way of a cause, the principles from which a syllogism proceeds must also be necessary and thus cannot be otherwise than they are. For a necessary effect cannot come from a non-necessary cause" (*In Meta.* V, 6.836-838).

However, given a different set of conditions, necessity can be found always or for the most part. In other words, if certain conditions change, a necessary condition is predominantly the case. For example, although it is necessarily the case that man is a rational animal that walks on two legs, not all men walk on two legs and if in a vegetative state, they may not be rational.

By returning to the sensible conditions of human knowledge, a universal and necessary causal relationship can be applied to any efficient cause and effect occurrence that obtain always and everywhere and therefore necessarily, or that necessarily obtain always or for the most part. Therefore, *one obtains universal and necessary causality through abstraction of singular efficient causal events that are found to occur always and everywhere or for the most part as one reflects upon these singular events through the vis cogitativa. Once a universal and necessary causal relationship is determined to exist by an act of the intellect in forming a judgment where the cause and its effect are reunited, that universal and necessary causal relationship judged to exist by the intellect can be reapplied through the cogitative power to any singular efficient causal relationship in existence to determine its necessity and universality.* In this way one has knowledge of synthetic *a priori* knowledge of efficient causality independent of the transcendental philosophy. Again independent of the transcendental philosophy in the sense of being independent of the rules for the imaginative synthesis of the manifold or simply independent of the rules or schemata some of which are empirical and others are *a priori*. The following section will discus further how efficient causal intentions are apprehended by the intellect as universal and necessary.

Incidental Sensibles and a priori Knowledge of Efficient Causality in Aquinas

Aquinas explains that a special sensible object is that which is perceived by only one sense and no other and in respect of this special sensible object the perceiving sense cannot err. In the case of sight, color is its proper sensible.[324] In the case of hearing, sound; in the case of taste, flavor, and smell, odor.[325] Touch, has several special or proper sensible objects "proper to itself: heat and moisture, cold and dryness, the heavy and the light, etc. Each sense judges the objects proper to itself and is not mistaken about these, e.g. sight with regard to such and such a color or hearing with regard to sound."[326]

Aquinas also explains, that although the external senses in apprehending a sensible object is not numerically the same as that object, yet it may be similar insofar as it is affected by that object. [327] There are cases where forms are apprehended by the apprehensive powers and the senses are made like their sense objects insofar as physical changes take place similar to that of the sensible object (e.g., the hand becomes cold in touching a cold stone), but there is no material alteration on the part of the senses whereby the senses are made identical to the sensible objects such that the hand in becoming cold becomes the stone that is touched.[328]

There are also cases where no physical alteration takes place in apprehending an intention not directly attainable from any sensible object, but only indirectly apprehended from sensibility.[329] This would be the case with incidental sensibles for example where no direct alteration of the external senses take place when one apprehends an efficient causal relation between a given cause and its effect. However, in the case of *per se* sensibles, a physical alteration does take place such as heat and cold objects making the sense of touch hot or cold; or, in the case of sight absorption of photons of particular wavelengths change the shape of pigments in the case of rod or cones within the retina.

Following Aristotle, Aquinas maintained that the senses are not deceived in relation to their proper sensible objects but can be "deceived

---

324 *In De Anima* II 13.384.
325 Ibid.
326 Ibid.
327 *In De Anima* II 24.551-552.
328 *In De Anima* II 24.553-554
329 *In De Anima* II 24.553.

both about objects only incidentally sensible and about objects common to several senses (*sensibilia communia*). Thus sight would prove fallible were one to attempt to judge by sight what a coloured thing was or where it was; and hearing likewise if one tried to determine by hearing alone what was causing a sound. Such then are the special objects of each sense."[330] The senses can thus be deceived in either the case of incidental sensibles or in the case of common sensibles where common sensibles are such that the individual judges by sight what a colored thing is and where it is, or from hearing, what was causing the sound.[331] In both cases, more than one sense is required and judgment is needed to make the determination.

Aquinas lists the various common sensibles that are not proper to one sense but common to all senses insofar as the various senses can be used to distinguish between these various common sensibles. Aquinas explains, "the common sense-objects are five: movement, rest, number, shape and size. These are not proper to any one sense but are common to all; which we must not take to mean that all these are common to all the senses, but that some of them, i.e. number, movement and rest, are common to all. But touch and sight perceive all five".[332] Aquinas lists five common sensibles consisting of movement, rest, number, shape and size.

Galileo designated these common sensibles as primary sensibles qualities and designated Aristotle's proper sensibles as secondary qualities believing that extension or number, size, and shape do exist in external objects and are measurable whereas the proper sensibles only exist as perceived.[333] Neither the proper nor the common sensibles

---

330 *In De Anima* II 13.385.
331 Ibid.
332 *In De Anima* II 13.386.
333 Galileo in The Assayer explains "I do not at all feel myself compelled to conceive of bodies as necessarily conjoined with such further conditions as being red or white, bitter or sweet, having sound or being mute, or possessing a pleasant or unpleasant fragrance. On the contrary, were they not escorted by our physical senses, perhaps neither reason nor understanding would ever, by themselves, arrive at such notions. I think, therefore, that these taste, odors, colors, etc., so far as their objective existence is concerned, are nothing but mere names for something which resides exclusively in our sensitive body (*corpo sensitivo*), so that if the perceiving creatures were removed, all of these qualities would

be annihilated and abolished from existence. But just because we have given special names to these qualities, different from the names we have given to the primary and real properties, we are tempted into believing that the former truly exist as well as the latter" (Galileo's *Il Saggitore* (The Assayer), which appeared originally in 1623. Translation by A. C. Danto, from *Introduction to Contemporary Civilization in the West* 2nd ed. vol. I (New York: Columbia University Press, 1954), 719-24).

Galileo makes clear that taste, colors, odors, or other "secondary qualities" are not external and are instead mind-dependent or perception dependent and merely names. However, the primary qualities exist in external bodies: "I cannot believe that there exists in external bodies anything, other than their size, shape, or motion (slow or rapid), which could excite in us our tastes, sounds, and odors. And indeed I judge that, if ears, tongues, and noses be taken away, the number, shape, and motion of bodies would remain, but not their taste, sounds, and odors. The latter, external to the living creature, I believe to be nothing but mere names, just as (a few lines back) I asserted tickling and titillation to be, if [the sense of touch be] removed" (Ibid.). Likewise, John Locke's representational realism held that only primary qualities such as extension or size, shape, motion or rest, number, and solidity are primary qualities of an external object and those which are measurable. Color and other secondary qualities were not considered to be an intrinsic property of an object. The object has the power to produce color in the right conditions, if given the right lighting conditions for an observer. Tables are a number of colorless subatomic particles moving about in space. If one closes their eyes or walks about a dark room, the color of a table disappears. However, one can still feel the shape, weight, etc. of the table and so our perception can represent certain primary sensible qualities of the object as the thing is in itself for Locke, but secondary sensible qualities are merely relative to the person perceiving them and not qualities in external objects.

Berkeley took Galileo and Locke's representative realism one step further and argued that an object's primary qualities are likewise relative to perception and observer dependent just as are secondary qualities. So the size of an object is relative to the distance of an object, shape relative to the angle from which it is viewed, weight relative to the strength of the individual. Berkeley thus concluded that if all we know of an external object is our perception of the object then one can never tell if such perceptions accurately represent the object. Therefore we can never perceive the object as it is in itself, or perceive what the object is as it is. For Berkeley, to be is to perceive or to be perceived, and nothing can exist unperceived. In *A New Theory of Vision* Berkeley concluded that we do not directly perceive distance but infer it, and in effect, we know only ideas or perceptions and not the objects of experience leaving only a "veil of perception." Hume and Kant inherited this legacy from early modern philosophy. However, one must ask, do chemical compounds within an external object exist and if not then how is chemistry even possible? If color does not exist in an object, how can that object possibly have a power by which perception perceives color? If size does not exist outside perception, then how is size measurable and mathematics possible? If a table is merely atoms in motion, then how could I possibly experience or perceive the width, breadth, and size of a table much less touch and experience solidity?

Galileo, Locke, and Berkeley were partially correct to suggest that perception does indeed interpret experience in certain cases, but they were mistaken to assume that secondary or primary qualities are not in or related to a given external subject. One who is color blind cannot see the color red, and one may perceive a distant object varying in size and shape. However, one can judge that a distant object is not the size of a pen but the size of a house and find that in fact the distant shape is the size of a house being in fact a house. Under normal conditions of visual perception individuals share in the same experience of color as this blue sky, or this red apple. Two different individuals can measure the same width, length, and breadth and find the same identical dimensions of a table without fear that the measurement will change from moment to moment. Likewise, two individuals hear the same sound and share in the same musical tone, rhythm, melody, and beat making music possible. This suggests that external sensible qualities that are secondary and primary, proper and common, can all likewise be commonly perceived as numerically one and the same in a given subject and be perceived as individuated according to one's faculties of perception and external senses allow. Thus the "veil of appearance" is removed and one can apprehend through experience those primary and secondary qualities that indicate the real essence of

include efficient causality or substance. Following Aristotle, number, movement, and rest were considered common to all the senses but shape and size were perceived not by every sense but by specific senses. For example, size or shape might be perceived by sight and touch but not perceived by taste or smell. Taste can perceive distinct flavors only, but touch can distinguishing motion, rest, or extension.

Aquinas then discusses incidental sensibles that are different from proper or common sensibles. For example, both Diarus and Socrates might be some color, but Diarus and Socrates are only perceived incidental to the perception of a particular color and various features. The color red, black, or white is a proper *per se* sensible and thus sensed directly by sight, but in sensing the sense-object one judges that this sense-object is Diarus or Socrates. In sensing the red colored individual, one incidentally perceives that the red object is this or that person. Aquinas commenting on Aristotle explains: "we might, he says, call Diarus or Socrates incidentally a sense-object because each happens to be white: that is sensed incidentally (*sentitur per accidens*) which happens to belong to what is sensed absolutely (*sentitur per se*). It is accidental to the white thing, which is sensed absolutely, that it should be Diarus; hence, Diarus is a sense-object incidentally. He does not, as such, act upon the sense at all."[334]

The son of Diarus is an incidental object of sense or *intentiones non-sensatae* or *intentiones insensatae*. As Anthony J. Lisska explains, "these are acts of awareness to which there is not an object in the external world that acts causally on the sense faculty."[335] Aquinas writes, that these are neither "immediately nor directly (*nec primo nec per se, sed per accidens*), as when the likeness [*similitudo*] of a person is in the sight.

what this or that thing is. Further, our perception is not of perception or of sense data, instead Aquinas argues that our experience is of the things themselves and what is apprehended by the intellect are real essences and not merely nominal essences. Therefore, we are assimilated to intentional forms of external objects, which are the things themselves and not assimilated to perceived appearances. Thus the transcendental philosophy itself is no longer needed, nor does it accurately reflect sense experience or that of perception.

334 *In De Anima* II 13.387.

335 Anthony J. Lisska, "Thomas Aquinas on Phantasia: Rooted in But Transcending Aristotle's De Anima", (www.nd.edu/Departments/Maritain/ti00/lisska.htm), 5.

She is not there [i.e., in the sight *per se*] because she is a human person, but because she is a colored object;"[336] The *intentiones insensatae* "are neither identical with nor co-extensive with the mental act of concept-formation, which, of course, is what Aquinas identifies with the role of abstraction through the *intellectus agens*."[337] To have the quiddity is to be intentionally assimilated to the "nature" or "essence" of a thing. The *vis cogitativa* according to Lisska, "'interprets' an object as an individual of a kind and not merely as a unified concrete whole composed of a set of proper and common sensibles. This, in effect, distinguishes the *vis cogitativa* from the *sensus communis*."[338]

Aquinas also explains that although common and special sensibles are *per se* sensibles (*per se sensibilia*), or perceived by the senses, it is the proper *per se* sensibles (*proprie per se sensibilia*) that are directly perceived.[339] Motion for example is not directly perceived strictly speaking in the same way color is perceived by sight. Sight perceives the color of something in motion and through the color being in motion, motion is perceived. Aquinas goes further and explains that the internal sense called the common sense faculty is where the modifications affecting all the particular senses terminate. This common sense faculty in whatever sense it may be located in or throughout the brain becomes aware of modifications to the particular external senses at the level of sense perception, and is aware of the differences between these senses. By means of the common sense one is not only aware of one's own life (instinctively aware of self-motion for example, or hunger, or sadness), but also the distinction between whiteness and sweetness of a given sensible object.[340] Thus by means of the common sense, one is able to "distinguish between the objects of different senses."[341]

---

336 *ST* I q17, a2
337 Anthony J. Lisska, "Thomas Aquinas on Phantasia: Rooted in But Transcending Aristotle's De Anima", (www.nd.edu/Departments/Maritain/ti00/lisska.htm), 9.
338 Ibid., 9.
339 *In De Anima* II 13.387.
340 *In De Anima* II, 13.390.
341 *In De Anima* II, 13.390.

Aquinas explains further that "sensation is a being acted upon and altered in some way. Whatever, then, affects the faculty in, and so makes a difference to, its own proper reaction and modification has an intrinsic relation to that faculty and can be called a sense-object in itself or absolutely. But whatever makes no difference to the immediate modification of the faculty we call an incidental sense-object. Hence, the Philosopher says explicitly that the senses are not affected at all by the incidental object as such."[342] Sensation is being acted upon and altered by some sensible object, affecting the faculty and causing a reaction and modification to that faculty.

In addition, the immediate objects of sensation differentiate sense-experience so that precisely by their differences is sensation itself differentiated since the various kinds of stimulants of sensation are, in their actuality as such, precisely the special sense-objects.[343] For example, the color red distinguishes between the sensation of red and black in the case of sight, or between the sensation of heat and cold in the case of touch since the sensation of red or heat are the special sense-objects themselves acting upon the external senses. However, whatever does not make a direct or immediate modification to a faculty of sense is an incidental sensible and thus the senses are not affected by the incidental object as such. This is the case with efficient causality since efficient causality does not immediately affect or modify a given external sense or a given sense faculty by acting upon it.

Further, various sense objects affect the senses in different ways, for example if they are large or small, near or far, together or apart, and thus size and position of objects very between the five senses.[344] However, this mode of stimulating rather than being a kind of stimulation does not immediately act upon the senses therefore they do not differentiate the sense-faculties but remain common to several faculties at once.[345] Having explained the distinction between proper

---

342 *In De Anima* II, 13.393.
343 *In De Anima* II, 13.394.
344 *In De Anima* II, 13.390.
345 *In De Anima* II, 13.390.

and common sensibles and the difference between kind and mode of sensation, Aquinas is ready to explain how a sense-object can be considered 'incidental.'[346]

First, for a sense object to be considered incidental it must be connected accidentally with a *per se* common or proper sensible.[347] For example, a man may be accidentally white, or a white thing may be accidentally sweet.[348] Second, it must be perceived by the one sensing and thus known by means of some cognitive faculty where the cognitive faculty knows its sensible object *per se*.[349] Another sense faculty, or the intellect, or the cogitative faculty (*vis cogitativa*) as in the case of humans, or natural instinct (*vis aestimativa*) as in the case of other animals must know a sensible object *per se* to perceive the incidental sensible.[350] For example, white is a *per se* proper sensible as is a sweet taste, but saying this white thing is sweat is incidental to it being white. To perceive that this white object is sweet one must first sense the color and also the taste of sweetness prior to being able to apprehend the incidental relationship between some white object itself also being sweet. To incidentally apprehend that a particular white object itself is sweet prior to tasting the object, the *vis cogitativa* recollects from memory, experience, and imagination that similar white objects were sweet.

The cogitative faculty from these *per se* proper sensibles, seeing white and tasting something sweet, judges that a white thing that is seen is in fact sweet. As one perceives a white object, the cogitative power through reminiscence recalls similar white objects that were sweet and judges incidentally that a given white object is sweet. The incidental association between sweetness and a particular white object is incidentally perceived by the cogitative power, by instinct, or by some other sense-faculty.[351] Thus incidental sensibles can be perceived by a

---

346 *In De Anima* II, 13.395.
347 *In De Anima* II, 13.390.
348 *In De Anima* II, 13.390.
349 *In De Anima* II, 13.390.
350 *In De Anima* II, 13.390.
351 Aquinas explains: "Having seen how we should speak of the absolute or essential sense-objects, both common and special, it remains to be seen how anything is a sense-object 'incidentally'. Now for

number of different faculties. Aquinas explains, however, this is strictly speaking not fully an incidental sensible since it remains essentially sensible.[352] Aquinas then makes a distinction between incidentals known by the intellect itself and those known by the *vis cogitativa*:

> What is not perceived by any special sense is known by the intellect, if it be a universal; yet not anything knowable by intellect in sensible matter should be called a sense-object incidentally, but only what is at once intellectually apprehended as soon as a sense-experience occurs. Thus, as soon as I see anyone talking or moving himself my intellect tells me that he is alive; and I can say that I see him live. But if this apprehension is of something individual, as when, seeing this particular colored thing, I perceive this particular man or beast, then the cogitative faculty (in the case of man at least) is at work, the power which is also called the 'particular reason' because it correlates individualized notions, just as the 'universal reason' correlates universal ideas.[353]

---

an object to be a sense-object incidentally it must first be connected accidentally with an essential sense-object; as a man, for instance, may happen to be white, or a white thing happen to be sweet. Secondly, it must be perceived by the one who is sensing; if it were connected with the sense-object without itself being perceived, it could not be said to be sensed incidentally. But this implies that with respect to some cognitive faculty of the one sensing it, it is known, not incidentally, but absolutely. Now this latter faculty must be either another sense-faculty, or the intellect, or the cogitative faculty [*vis cogitativa*], or natural instinct [*vis aestimativa*]. I say 'another sense-faculty', meaning that sweetness is incidentally visible inasmuch as a white thing seen is in fact sweet, the sweetness being directly perceptible by another sense, i.e. taste. But, speaking precisely, this is not in the fullest sense an incidental sense-object; it is incidental to the sense of sight, but it is essentially sensible. Now what is not perceived by any special sense is known by the intellect, if it be a universal; yet not anything knowable by intellect in sensible matter should be called a sense-object incidentally, but only what is at once intellectually apprehended as soon as a sense-experience occurs. Thus, as soon as I see anyone talking or moving himself my intellect tells me that he is alive; and I can say that I see him live. But if this apprehension is of something individual, as when, seeing this particular colored thing, I perceive this particular man or beast, then the cogitative faculty (in the case of man at least) is at work, the power which is also called the 'particular reason' because it correlates individualized notions, just as the 'universal reason' correlates universal ideas. Nevertheless, this faculty belongs to sensitivity; for the sensitive power at its highest--in man, in whom sensitivity is joined to intelligence--has some share in the life of intellect" (*In De Anima*, II, Lec. 13, Sct. 395-397; emphasis added).

352 *In De Anima* II, 13.396.
353 *In De Anima* II, 13.396.

The distinction is made between what is apprehended by the intellect as a universal at the moment that sense-experience occurs and what is apprehended of this or that particular by the cogitative power. An incidental sensible or a determinate nature, grasped by the cogitative power is not strictly speaking a universal since it is of the singular and sensible, but in another way "sense is in a certain way even of the universal."[354] The intellect apprehends at the moment of sense-experience a universal such as "living" by seeing one move, or talk. In the case of particulars, what is apprehended by the cogitative power from seeing this particular colored thing, is the perception that it is this or that particular individual Socrates or Diarus, but the intellect knows directly in the universal Socrates, Darius, friend, or living when the sense indirectly perceives these indirect objects of sense. From seeing particular colors, the cogitative faculty distinguishes or correlates incidental sensibles of Darius or friend insofar as colors belong to this particular man or beast. Aquinas explains this relationship further:

> A thing is perceptible to the senses of the body in two ways, directly and indirectly. A thing is perceptible directly if it can act directly on the bodily senses. And a thing can act directly either on sense as such or on a particular sense as such. That which acts directly in this second way on a sense is called a proper sensible, for instance color in relation to the sight, and sound in relation to the hearing. But as sense as such makes use of a bodily organ, nothing can be received therein except corporeally, since whatever is received into a thing is therein after the mode of the recipient. Hence all sensibles act on the sense as such, according to their magnitude: and consequently magnitude and all its consequences, such as movement, rest, number, and the like, are called common sensibles, and yet they are direct objects of sense.

---

354 *In Post. Anal.* II, 20 (*Exposition on the Posterior Analytics of Aristotle.* Translated by Pierre H. Conway. Quebec: *La Libraire Philosophique* M. Doyon, 1956; also see Alfred Leo White, Jr. *The Experience of Individual Objects In Aquinas A Dissertation* (Washington, D.C.: The Catholic University of America, 1997), 210-211.

An <u>indirect object of sense is that which does not act</u> <u>on the sense, neither as sense nor as a particular sense, but is</u> <u>annexed to those things that act on sense directly: for instance</u> <u>Socrates; the son of Diares; a friend and the like which are</u> <u>the direct object of the intellect's knowledge in the universal,</u> <u>and in the particular are the object of the cogitative power</u> <u>in man, and of the estimative power in other animals.</u> The external sense is said to perceive things of this kind, although indirectly, <u>when the apprehensive power (whose province it is</u> <u>to know directly this thing known), from that which is sensed</u> <u>directly, apprehends them at once and without any doubt</u> <u>or discourse (thus we see that a person is alive from the fact</u> <u>that he speaks): otherwise the sense is not said to perceive it</u> <u>even indirectly.</u>[355]

Aquinas makes a distinction between what is perceived directly and indirectly by the senses. What is known directly are proper and common sensibles where common sensibles act upon the senses directly by means of magnitude and its consequences. For example, the extension of the table can be measured as can its motion across the room thus motion, number, and rest are directly perceived by the senses as and in relation to direct objects of sense.

The indirect objects of sense are those that do not directly act on the sense. These are not inherently part of sensation itself nor are they any particular sense. However, indirect objects of sense are *annexed* to things that directly act on sense. Aquinas gives several instances of incidental sensibles that are indirect objects of sense. For instance, Socrates, the son of Darius, a friend, or living, or one could add an efficient causal *connexion* where the *connexion* is *annexed* to a given cause and its effect where the events themselves are direct objects of sense but the *connexion* is an indirect object of sense.

---

355 *ST* III, q. 92, a. 2 body.

Aquinas then explains that these incidental intentions are "direct object(s) of the intellect's knowledge in the universal," and are at the same time, "in the particular," "the object of the cogitative power in man." Thus incidental sensibles such as the causal connexion if taken as a single instance is a direct object of intellectual knowledge in the universal while at the same time being perceived by the cogitative sense in the particular. However, in the case of repeated instances of causal incidental *connexions* annexed to contiguity and succession between causes and their effects, whether stored in memory and discursively judged to exist by the cognitive faculty (*vis cogitative*) through reminiscence and by returning to phantasms (or, to the external senses) in forming a disposition of custom, a universal *connexion* is formed by multiple instances of similar contiguous and successive instances of causal incidental *connexions*.

It is the particular reason that correlates individual intentions in forming experience and the universal from experience, just as the universal reason correlates universal forms.[356] It is the intellect that apprehends the universal and this is incidental to what is directly perceived by the external senses. Similarly, in the case of a single instance, as in the case of one seeing an individual talking or moving, the intellect apprehends immediately that the individual is alive; and one can thus say that he sees him live even though life itself is incidental to the perception of the individual talking or moving. Aquinas in the *De Anima* commentary then draws upon the Dionysian principle

---

356 Klubertanz indicates that "Averroës asserts that the sensible *per accidens* is apprehended by 'common sense' and 'distinguished' (that is, thought about; *distinguit == dianoei tai*) by the discursive power. It is therefore by reference to Averroës, not to Aristotle, that St. Thomas speaks here [*De Anima* II, 13] of his own theory of the discursive reason" (George P. Klubertanz, S.J., *The Discursive Power*, (Ohio: The Messenger Press, 1952), 198). Aquinas's development of the role of the discursive reason in judging *per accidens* sensibles provided the necessary bridge between sensibility and intelligibility of those sensibles not directly sensed by the external senses, but that are of the particular. Averroës explains: "He [Aristotle] did not mean that the sense comprehends the essences of things, as some thought; for this belongs to another power which is called intellect. But he meant that the senses, together with the comprehension of their proper sensibles, comprehend individual intentions distinct in genera and species. Therefore, they comprehend the intention of this individual man, and the intention of this individual horse, and generally the intention of the ten individual predicaments, and this seems to be proper to the senses of man...And this individual intention is that which the cogitative power distinguishes from the imagined form and strips it from the common and proper sensibles which are joined to it, and places it in the memorative power" (Ibid., 198).

of hierarchy to explain the connection between the universal and particular reason:

> Nevertheless, this faculty belongs to sensitivity; for the sensitive power at its highest--in man, in whom sensitivity is joined to intelligence--has some share in the life of intellect. But the lower animals' awareness of individualized notions is called natural instinct, which comes into play when a sheep, e.g., recognizes its offspring by sight, or sound, or something of that sort.[357]

Aquinas then makes clear that the cogitative particular reason is united to the universal intellect in one and the same subject allowing for particulars to be known in a common nature for instance Socrates or Diarus known as a man.[358] In the case of efficient causality, the cogitative power knows this or that incidental causal relation to be an instance of causality due to the universal intellect and the particular intellect being united in one and the same subject and operation. By seeing a given effect follow a given cause, the cogitative power is aware of the individualized intention or notion of an efficient causal relationship and when apprehended in a common nature, the cogitative power knows this or that efficient causal relationship to be causal. Early modern philosophy failed to see the single operation of the intellective power acting through the cognitive faculty in perception in the immediate apprehension of incidental sensibles such as the causal *connexion* between a given cause and its effect. [359]

---

357 *In De Anima* II, 13.397.

358 *In De Anima* III, 398.

359 The intellect is a blank tablet for Aquinas following Aristotle, thus first principles must be apprehended from experience and incidental sensibles. First principles such as the law of contradiction where "it is impossible for the same thing to be, and not to be" and the principle of identity where "two things cannot be numerically one and the same at the same time and in the same respect," or in the case of propositions that are self-evident in themselves but not necessarily self-evident to us unless we know the terms of the proposition as in the case of self-evident propositions with common terms such as "the whole is the sum of its parts" are all incidental known by the intellect and apprehended by the intellect from sensibility. John Locke, consistent with Aristotle and Aquinas argued that first principles are not innate, neither are axioms such as "the whole is the sum of its parts." Following a common empiricist tradition, Locke likewise argued with Aquinas and Aristotle that the intellect is a blank tablet on which nothing is written. See Locke *Echu* bk 1, chpt. III *Other considerations concerning Innate Principles, both Speculative and Practical*. Also, see *ST* I, q. 101, a. 1, body; *On Being and Essence* Chpt. 4 Para. 10.

Lisska points out "there is no analogue in classical British empiricism, for instance, for the incidental object of sense. Given the bundle view of perception espoused by Berkeley in *The Principles* and Hume in the *Inquiry*, among other places, theoretically there is

Aquinas in regard to first principles makes a distinction between what is self-evident in itself and what is only self-evident upon understanding the meaning of the terms of a given proposition: "It must be noted that self-evident propositions are those which are known as soon as their terms are known, as is stated in Book I of the Posterior Analytics (*Analytica Posteriora*, I, 3 (72b 18)). This occurs in the case of those propositions in which the predicate is given in the definition of the subject, or is the same as the subject. But it happens that one kind of proposition, even though it is self-evident in itself, is still not self-evident to all, i.e., to those who are ignorant of the definition of both the subject and the predicate. Hence Boethius says in the Hebdomads (*De Hebdomadibus* (i.e., *Quomodo Substantiae*) (PL 64, 1311)) that there are some propositions which are self-evident to the learned but not to all. Now those are self-evident to all whose terms are comprehended by all. And common principles are of this kind, because our knowledge proceeds from common principles to proper ones, as is said in Book I of the *Physics* (*Physica*, I, 1 (184a 21)). Hence those propositions which are composed of such common terms as whole and part (for example, every whole is greater than one of its parts) and of such terms as equal and unequal (for example, things equal to one and the same thing are equal to each other), constitute the first principles of demonstration. And the same is true of similar terms. Now since common terms of this kind belong to the consideration of the philosopher, then it follows that these principles also fall within his scope. But the philosopher does not establish the truth of these principles by way of demonstration, but by considering the meaning of their terms. For example, he considers what a whole is and what a part is; and the same applies to the rest. And when the meaning of these terms becomes known, it follows that the truth of the above-mentioned principles becomes evident" (*In Meta*. Bk 4, Lsn. 5, Sct. 325; emphasis added). Aquinas likewise argues that all axioms or demonstrations must be reduced to the first principle of non-contradiction or contradiction for the law of non-contradiction is the principle that is best known and not hypothetical since something *cannot* both be and not be at the same time and in the same respect. Cf. *In Meta*. Bk. 4, Lsn. 6, Sct. 326-328.

Aquinas also explains that first principles are apprehended immediately by understanding and are not arrived at through discursive reason, "For the act of "understanding" implies the simple acceptation of something; whence we say that we understand first principles, which are known of themselves without any comparison. But to "reason," properly speaking, is to come from one thing to the knowledge of another: wherefore, properly speaking, we reason about conclusions, which are known from the principles… But it has been shown above (*ST* I q. 79, a. 8) that it belongs to the same power both to understand and to reason, …." (*ST* I, q. 83, a. 4 body). In addition, Aquinas describes the *inferior* apprehensive powers and their relation to the intellect explaining: "the intellect, of necessity, receives from the inferior apprehensive powers: wherefore if the imaginative, cogitative, or memorative powers be disturbed, the action of the intellect is, of necessity, disturbed also" (*ST* I, q. 115, a. 4 body).

Aquinas then explains the relationship between the cognitive apprehensive powers and first principles: "In the apprehensive powers, we must observe that there are two passive principles: one is the "possible" [cf. *ST* I, q. 79, a. 2 ad 2] intellect itself; the other is the intellect which Aristotle (*De Anima* iii, text 20) calls "passive," and is the "particular reason," that is the cogitative power, with memory and imagination. With regard then to the former passive principle, it is possible for a certain active principle to entirely overcome, by one act, the power of its passive principle: thus one self-evident proposition convinces the intellect, so that it gives a firm assent to the conclusion, but a probable proposition cannot do this. Wherefore a habit of opinion needs to be caused by many acts of the reason, even on the part of the "possible" intellect: whereas a habit of science can be caused by a single act of the reason, so far as the "possible" intellect is concerned. But with regard to the lower apprehensive powers, the same acts need to be repeated many times for anything to be firmly impressed on the memory. And so the Philosopher says (*De Memor. et Remin.* 1) that "meditation strengthens memory"" (*ST* I-II, q. 51, a. 3).

no room left for the incidental object of sense. Berkeley and Hume analyze an individual in terms of a collection of sensible properties."[360] As indicated by Lisska, Berkeley wrote in *The Principles*: "Thus, for example, a certain color, taste, smell, figure and constancy, having been observed to go together, are accounted one distinct thing, signified by the name 'apple.' Other collections of ideas constitute a stone, a tree, a book, and the like sensibles things."[361]

Hume following Berkeley wrote in *An Inquiry Concerning Human Understanding*: "As our idea of any body, a peach, for instance, is only that of a particular taste, color, figure, size, consistency, etc., so our idea of any mind is only that of particular perceptions without the notion of anything we call substance, either simple or compound."[362] Hume likewise following the nominalism of Galileo and the British Empiricists maintained that the idea of substance is "nothing but a collection of Simple ideas, that are united by the imagination, and have a particular name assigned them, by which we are able to recall, either to ourselves or others, that collection."[363]

---

360 Anthony J. Lisska, "Thomas Aquinas on Phantasia: Rooted in But Transcending Aristotle's De Anima", (www.nd.edu/Departments/Maritain/ti00/lisska.htm), 10.

361 Ibid.; George Berkeley, *A Treatise Concerning the Principles of Human Understanding*, #1.

362 David Hume, *An Inquiry Concerning Human Understanding* (Hendel Edition), 194.

363 David Hume, *A Treatise of Human Nature* I, Part 1, Sec. 6 (Selby-Biggs edition). Likewise Hume explains "all general ideas are nothing but particular ones, annexed to a certain term, which gives them a more extensive signification, and makes them recall upon occasion other individuals, which are similar to them" (*A Treatise of Human Nature* I, Part 1, Sec. 7). Although Berkeley opposed the doctrine of abstraction, Hume's nominalism and theory of abstraction follows essentially the nominalism and theory of abstraction of Berkeley. Berkeley explains his theory of abstraction: "an idea which considered in itself is particular becomes general by being made to represent or stand for all other particular ideas of the same sort" (*Principles of Human Knowledge*, Introduction, §12). Likewise Locke although giving an account of abstraction remained a nominalist: "the same colour being observed to day in Chalk or Snow, which the mind yesterday received from Milk, it considers that appearance alone, makes it a representative of all that kind; and having given it the name Whiteness it by that sound signifies the same quality wheresoever to be imagin'd or met with; and thus Universals... are made" (*Essays Concerning Human Understanding*, II, xi, 9). Again, rather than abstracting common matter such as bone and flesh from this or that individual, Locke abstracts similarity (extensionality rather than the essence) as in a child arriving at the general idea of man: "that there are a great many other things in the World, that in some common agreements of Shape, and several other Qualities, resemble their Father and Mother, and those Persons they have been used to... wherein they make nothing new, but only leave out of the complex idea they had of Peter and James, Mary and Jane, that which is peculiar to each, and retain only what is common to them all" (*Essays Concerning Human Understanding*, III, iii, 7). Michael Ayers draws the relationship between Locke and Berkeley see Michael Ayers, *Locke* (London: Routledge, 1991), I 250-251). Hence Locke, Berkeley, and Hume shared in a similar theory of abstraction based

Lisska explains that this "paradigm of perception" is considered the bundle theory view where physical objects are nothing more than a bundle of sensible qualities acting as primary and secondary qualities.[364] However, Lisska indicates further that,

> [Aquinas] goes beyond the limits of the bundle view paradigm. The *vis cogitativa* is what provides for this more sophisticated account... Aquinas, in suggesting that it is by means of the *vis cogitativa* that we account for our perception of the incidental object of sense, goes beyond the limits found in Aristotle's account of sensation and perception... both Aristotle and Aquinas transcend the limits entailed by the bundle view paradigm adopted by the British empiricists.... In effect, inner sense goes beyond the data of the external sensorium. This is accomplished through an *intentio non-sensata* and the structured mental act of the *vis cogitativa*... the *vis cogitativa* explains the possibility of the awareness of individual substances.[365]

Lisska indicates that the intellect apprehends substance through the *vis cogitativa* and the *intentio non-sensata*. If what is apprehended is the composite of substance and accidents then there is no reason for the intellect not to apprehend this composite in apprehending a given subject, but just as in seeing this or that colored object one apprehends the incidental sensible 'the son of Darius' it seems to be the case that

---

in nominalism even if Hume's theory does in fact account for empirical generality through custom and disposition. For Hume resemblance is the disposition to apprehend natural patterns of association such that the same word calls to mind those things that are similar and yet is able to distinguish from those things that are different (see Hanna Ginsborg, "Thinking the Particular as Contained under the Universal," Dept. of Philosophy, U. C. Berkeley Draft, (November 2003), 16-17. Also see Wilfred Sellars, "Empiricism and the Philosophy of Mind," in Science, Perception and Reality (London: Routldge and Kegan Paul, 1963), 160-161 where Sellars argues that one can arrive at "psychological nominalism" by "modifying" Hume's view. The position of this thesis is that psychological nominalism is in fact a legacy of the late Medieval nominalism influencing early modern philosophy and the new science and any modification to Hume's nominalism is only necessary to adapt early modern nominalism to contemporary nominalism.

364 Anthony J. Lisska, "Thomas Aquinas on Phantasia: Rooted in But Transcending Aristotle's De Anima", (www.nd.edu/Departments/Maritain/ti00/lisska.htm), 10.
365 Ibid., 10.

both causality and substance are apprehended through *intentio non-sensata* and the mental act of the *vis cogitativa*. Lisska continues:

> The possibility of our being aware of individual things is accounted for by means of the phantasm-structured *vis cogitativa*. The external sensorium is aware of unified wholes of proper and common sensibles. At this point in the process—i.e., the external sensorium—Aquinas's account is similar structurally to the bundle view paradigm articulated by Berkeley and Hume. The *vis cogitativa*, however, is aware of the primary substance as a primary substance—an individual. The mental act of the *vis cogitativa* renders the awareness of 'unified whole' into an awareness of an individual ... it is because of the *vis cogitativa* that Aquinas can distinguish between sensation and perception, and, a fortiori, transcend the limits of modern and contemporary British empiricism.... Through his analysis of the vis cogitativa, Aquinas undercuts the sense data theories of early twentieth century epistemology found in the writings of...the early modern empiricism of Locke, Berkeley, and Hume. Aquinas accomplishes this by suggesting, in effect, that our experience is of things rather than of sense data... Aquinas provides the philosophy of mind machinery necessary to explain the possibility of an act of awareness of an object beyond the immediate data of the proper and the common sensibles.[366]

A causal relationship of individual cause and effect events cannot be directly or immediately sensed as a *per se* sensible as Hume came to realize from the debates between the British Empiricists and Rationalists. However, what Kant following Hume failed to realize is that perception of incidental intentions or apprehension of incidental sensibles by the cognitive act rather than perception of *per se* sensibles provides the bridge between the particular and universal in the case of

---

366 Ibid., 11.

efficient causality, between the sensible and intelligible when it comes to causal relationships that are not *per se* sensible. Causal relationships of particular cause and effect events cannot be directly or immediately sensed as *per se* sensibles, but the causal relationship can be apprehended by the cogitative power in the case of an incidental sensible of the particular events, while causality itself can be apprehended as universal by the intellect.

For Aquinas, custom is a "predisposition in the memory to recall objects in a certain order, enables humans to reminisce about individuals perceived in the past according to their proximity in time and location to each other."[367] Custom provides a predisposition for apprehending spatio-temporal cause and effect relationships that are not directly sensible, but it is not custom itself that apprehends incidental sensibles since it is simply a predisposition. For spatio-temporal cause and effect relationships to be cognized, a cognitive power is required and it is this that Hume failed to realize. Since it is the cogitative power that apprehends continuity in time and location, it is the cogitative faculty that makes possible the predisposition of custom in reminiscence as well as apprehension of cause and effect relationships that are predisposed to be known by custom.[368] The cogitative power is that by which one

---

367 Alfred Leo White, Jr. *The Experience of Individual Objects In Aquinas A Dissertation* (Washington, D.C.: The Catholic University of America, 1997), 197; also see *In De memoria et reminiscencia*, chpt. 5.
368 A recent study in 2003 by Pierre Fonlupt provides evidence that particular brain regions are involved in judgment of causality and movement judgments (cf. Pierre Fonlupt, "Perception and judgment of physical causality involve different brain structures," *Cognitive Brain Research* vol. 17/2 (Nov. 2003), 248-254. The study investigated brain regions that exhibit increased activity during the judgment of causality (i.e., 'judged causality' where one judges the presence or absence of a causal relationship between balls) as compared with judgments that determine the direction of motion of the ball (i.e., 'perceived causality' or judgment of movement). For Aquinas this would be equivalent to a difference between causal incidental intentions and common sensible intentions of motion.

In the experiment, launching displays were presented to subjects who were instructed to respond to whether either a ball caused the movement of another ball (causality judgment) or if their was simply a direction of movement without causality (movement judgment). In the case of causation, a blue ball rolls horizontally across the screen and after 1 sec. collides with a red ball, and the red ball moves horizontally across the screen after having been struck by the blue ball. A non-causal event is where a blue ball passes underneath the red ball without striking the ball. Finally, a visual-transient experiment was conducted where a ball rolls across the screen and turns blue. The experiments were able to distinguish between the brain activations linked to causal perception and causal interpretation showing an increase of "medial frontal cortex activity when subjects were explicitly instructed to search for causality relative to when they were instructed to focus on the direction of movement" (Ibid., 252-253). This increase in medial

reminisces according to custom about a cause and effect relationship perceived in the past according to its proximity in time and location to another reoccurring cause and effect relationship, or between a given cause and its effect. In addition, it is by the cognitive power or the *vis cogitativa* distinguishing between a given cause and its effect that one perceives an individual incidental causal relationships from one or many instances of congruity similar to how first principles are formed from experience and repeated particular instances forming one common item consolidated in the intellect.[369]

---

frontal cortex activity was "specifically associated with the search for causality and not the perception of causality because the signal increase occurs whatever the nature of the stimulus (caused or non causal)" (Ibid., 253). Specifically, the maximum activity was found in the medial and dorsal part of the superior frontal cortex and involved "focused attention, memory, and reasoning" (Ibid., 253).

The experiment indicates that not only is the "visual system wired to recover the causal structure of the world," but also an interaction exists between "perception of causality and higher level processes that are used to interpret causality" (Ibid., 251, 253). Further, it has been shown that "perceived causality in a single launch event and the judgment of causality in the overall set can be differently affected by the nature of stimulus, like contiguity and contingency" (Ibid., 252). The judgment of causality can therefore be affected differently by stimulus that is either necessary, taken as contiguous, or contingent. The research concludes that a determination of physical causality involves a comparison between "a knowledge of physical laws that govern collisions" (laws of universal and necessary relationships), "the daily practice of these laws" (such as efficient causal events that are either necessary or contingent), and "the representation of a complex interrelation between objects" (Ibid., 253). For related research see A. Schlottmann, D.R. Shanks, "Evidence for a distinction between judged and perceived causality," *Q. J. Exp. Psychol.* 44 (1992), 321-342; S. Blakemore, P. Fonlupt, M. Pachot, C. Darmon, P. Boyer, A. Meltzoff, C. Segebarth, and J. Decety, "How the brain perceives causality: an event-related fMRI study," *Neuroreport* 12 (2001) 3741-3746; E. Castelli, F. Happe, U. Frith., and C. Frith., "Movement and mind: a functional imaging study of perception and interpretation of complex intentional movement patterns," *Neuroimage* 12 (2000) 314-325. For contemporary research in the cognitive basis for cogitative judgment see V. Goel, B. Gold, S. Kapur, and S. Houle, "The seat of reason? An imaging study of deductive and inductive reasoning," *Neuroreport* 8 (1997), 1305-1310; K. Christoff, V. Prabhakaran, J. Dorfman, Z. Zhao, J. Kroger, K. Holyoak, and J. Gabrieli, "Rostrolateral prefrontal cortex involvement in relational integration during reasoning," *Neuroimage* 14 (2001), 1136-1149.

369 Aquinas explains: "he [Aristotle] shows, in view of the foregoing, how the knowledge of first principles comes about in us; and he concludes from the foregoing that from sensing comes remembrance in those animals in which a sensible impression remains, as has been stated above. But from remembrance many times repeated in regard to the same item but in diverse singulars arises experience, because experience seems to be nothing else than to take something from many things retained in the memory.

However, experience requires some reasoning about the particulars, in that one is compared to another: and this is peculiar to reason. Thus, when one recalls that such a herb cured several men of fever, there is said to be experience that such a herb cures fevers. But reason does not stop at the experience gathered from particulars, but from many particulars in which it has been experienced, *it takes one common item which is consolidated in the mind and considers it without considering any of the singulars.* This common item reason takes as a principle of art and science" In Post. Anal. II, lec. 20; emphasis addeda).

The cogitative reminiscence according to custom of individual cause and effect relationships prepare the way for the intellect to abstract a common causal relationship or *connexion* and to distinguish between universal cause and effect relationships. It is by this abstraction of common or reoccurring causal relationship that the intellect apprehends the causal law. In the *Treatise of Human Nature*, the first question asked by Hume was how can it be said that every thing that has a beginning necessarily must have a cause, or simply *why is a cause always necessary*, or why is the causal maxim *whatever begins to exist, proceeds from some cause*, necessarily the case. Indeed why should one conclude a particular cause must necessarily have a particular effect, and what inference should be drawn from a particular cause to its effect?[370] Hume argued that impressions of causal relations can be distinguished separately and it is possible to conceive that a necessary connection between a cause and its effect does not in fact exist. Further, since the relation of causal ideas are not themselves unalterable as might be the case with quantity for example, there is no reason to hold that these relations are intuitively certain. Using these arguments Hume maintains that one can imagine fire without heat, or the bodies striking one another without motion; Hume *argues at least it is possible that they so occur*.

Hume correctly had seen the need for custom in apprehending cause and effect relationships. In a qualified sense Hume was also correct to realize that a necessary connection may not always exist between a given cause and its effect, and Hume was correct to see that incidental causal relationships may not always be properly distinguished or apprehended. If a cause is hindered from generating its effect there is no necessary relationship, but if an effect exists a cause must necessarily have produced the effect for the effect to exist suggesting absolute natural necessity in efficient causality. However, due to the early modern failure to account for incidental sensibles and the failure to see the significance of the cogitative power or cognitive faculties to distinguish

---

370 Anthony J. Lisska, "Thomas Aquinas on Phantasia: Rooted in But Transcending Aristotle's De Anima", (www.nd.edu/Departments/Maritain/ti00/lisska.htm), 27.

between individuals, Hume was unable to account for the relationship or necessary connection between a given cause and its effect.

Following Hume, Kant proposed the transcendental philosophy as an attempt to account for necessary cause and effect relationships associated with possible experience. To save science from Hume's critique on causality and substance, it was necessary for Kant to demonstrate a transcendental *a priori* basis for universal and necessary categories of substance and causality thereby demonstrating the causal law by appealing to the transcendental method. Early modern philosophy was unable to account for the necessary *connexion* between two contiguous events using primary qualities that Berkeley had relegated to ideas, and this similar to the way that Galileo had relegated secondary qualities to mere "names" following the nominalism that had influenced late Medieval philosophy.

Although Hume found that a necessary *connexion* is an essential element in causal relations, Hume argued that when one examines such relations one only finds contiguity and succession between a cause and its effect. Hume thus failed to realize that necessity in causality or simply a necessary *connexion* between a given cause and its effect could be apprehended through repeated occurrences of similar cause and effect incidental relationships or *connexions*. And similarly this incidental causal *connexion* could itself be apprehended from sensibility or experience using the imaginative, memorative, and cognitive faculties forming a disposition of custom. This in turn could allow for both belief in a causal inference and likewise the apprehension of a quasi-universal by the cognitive faculty or by the intellect itself as a universal incidental intention allowing for a necessary causal *connexion* to be known by the intellect. Either a single instance or a common occurrence of the incidental sensible *connexion* apprehended by the cogitative power once abstracted becomes the universal and necessary *connexion* between a given cause and its effect.

One can have this perception of the particular incidental sensible while at the same time apprehending the universal from the individual intention of efficient causality. Aquinas explains: "The cogitative faculty … apprehends the individual thing as existing in a common nature, and this because it is united to intellect in one and the same subject. Hence it is aware of a man as this man, and this tree as this tree; …"[371] The incidental efficient casual relationship is understood by the cogitative power as particular and by the intellect as universal being united in one and the same subject as an individual intention.[372]

Similarly, necessity can be apprehended from sensible efficient causal events that are necessarily related, and necessity can be apprehended to be universal through the process of abstraction from incidental cause and effect relationships that obtain always or in most cases. Thus cause and effect relationships can be apprehended of the particular as necessarily connected *always or for the most part* by the cogitative power and apprehended as both universal and necessary by the intellect when it is found that the causal relationship obtains either *everywhere and always, or always or for the most part if not hindered as in the case of contingent necessity.*[373] In opposition to Hume, however, the intellect apprehends

371 *In De Anima* II, 13.398.
372 Robert Pasnau indicates that "the term 'intention' is drawn from Avicenna's *Liber de anima* I.5 (85.88-86.6), where he distinguishes two kinds of sensory representations: forms and intentions. Whereas forms are apprehended by the five external senses and passed on to the inner sense, intentions are perceived only by inner sense, such as the cogitative power" (*Thomas Aquinas A Commentary on Aristotle's De Anima*, trans. by Robert Pasnau, (New Haven: Yale University, 1999), 208). Pasnau translates: "If, however, [the object] is apprehended as an individual—e.g., when I see something colored I perceived this human being or this animal—then this sort of apprehension in a human being is produced through the cogitative power. This is also called particular reason, because it joins individual intentions in the way that universal reason joins universal concepts (*rationum*). But all the same, this power is in the soul's sensory part. For the sensory power, at its highest level, participates somewhat in the intellective power in a human being, in whom sense is connected to intellect… the cogitative power apprehends an individual as existing under a common nature. It can do this insofar as it is united to the intellective power in the same subject. Thus it cognizes this human being as it is this human being, and this piece of wood as it is this piece of wood" (*De Anima* II, 13.396-398). The individual instances of a cause and effect connection is apprehended under a common nature by the cogitative power and this because the cogitative sensible feature of cognition is united to the intellective power in the same subject.
373 Aquinas writes: "all universals are said to be everywhere and always, in so far as universals are independent of place and time" (*ST* I, q16, a7, ad 2). Again, "Whence it is necessary that what is perceived by sense be this something, i.e., a singular substance, and that it be somewhere and now, i.e., in a determinate place and time. From which it is evident that what is universal cannot fall under the senses. For that which is universal is not determined to here and now, because then it would not be

inductively from incidental causal relationships the *intentio* that 1) every effect is connected to and proceeds from a cause, and 2) whatever begins to exist is an effect, then reason in the third operation of the intellect naturally deduces the causal maxim as a universal proposition from the major and minor premise. The basic syllogism might appear as follows:

> Every effect is connected to and proceeds from a cause
> Whatever begins to exist is an effect
> Whatever begins to exist is connected to and proceeds from a cause

The conclusion is the causal maxim, 'whatever begins to exist, proceeds from a cause'. Implicit in the causal maxim is what is indirectly or incidentally sensed which is the causal incidental *connexion* between a given cause and its effect. Likewise the incidental *connexion* implies that an effect proceeds from a cause and multiple instances implies that whatever begins to exist is an effect. If a *connexion* exists between a given cause and its effect and if the effect cannot exist without a cause then necessarily an effect proceeds from its cause. Likewise, if a *connexion* exists between a cause and its effect and if a cause must precede whatever comes to exist, then whatever begins to exist is an effect.

In the case of causality, the intellect abstracts from incidental sensibles that several causes produce various effects, and from this the intellect judges that these causes must produce their common effect by virtue of a higher cause to which the effects properly belongs. From the order of reason it is apprehended from the order of nature that *a proper effect is produced* by a particular cause *in respect of its proper nature or form*, thus it is understood that different causes having different natures and forms must have their respective different proper effects. In the order of reason and from the order of nature one judges that if different effects have a common cause everywhere and always, this is the proper effect of that cause and is necessary.

---

universal. For we say that is universal which is always and everywhere" (*In Posterior Analytics*, I, chpt. 42; all citations of *In Posterior Analytics* are from *Commentary on the Posterior Analytics of Aristotle*, trans. by F. R. Larcher (Albany, NY: Magi Books, 1970) unless otherwise noted).

Again in the order of reason and from the order of nature one judges that if different causes have a common effect, this is not the proper effect of any one of the causes, but of some higher cause by which they act. Aquinas gives an example from the order of nature that pepper, ginger, and the like differ in characteristics but they have a common effect of producing heat; but each has its particular effect differing from each of the others. In the order of reason, one traces the common effect of producing heat to a higher cause, specifically fire to which the common effect of producing heat belongs. It is the proper nature or form of fire to produce heat, and likewise the nature of different causes must have their different proper effects.

It is from incidental sensibles that one finds a common effect that pepper, ginger etc. all produce heat, and from this, one judges that some higher cause must exist to produce the common effect of heat. There may be repeated instances of this causal relationship, or contrary to Hume, simply one instance. In either case, it is the intellect at the order of reason that judges, through *per se* proper and common sensibles the incidental or *per accidens* sensible, that there is a common effect and a higher cause whereby pepper, ginger, wood, etc. produce heat by fire where fire is the higher cause of the common effect.[374]

---

374 Aquinas explains: "When several causes producing various effects produce one effect in common in addition to their various effects, they must needs produce this common effect by virtue of some higher cause to which this effect properly belongs. The reason for this is that since a proper effect is produced by a particular cause in respect of its proper nature or form, different causes having different natures and forms must needs have their respective different proper effects: so that if they have one effect in common, this is not the proper effect of any one of them, but of some higher cause by whose virtue they act: thus pepper, ginger and the like which differ in characteristics have the common effect of producing heat; yet each one has its peculiar effect differing from the effects of the others. Hence we must trace their common effect to a higher cause, namely fire to whom that effect properly belongs... Now all created causes have one common effect which is being, although each one has its peculiar effect whereby they are differentiated: thus heat makes a thing to be hot, and a builder gives being to a house. Accordingly they have this in common that they cause being, but they differ in that fire causes fire, and a builder causes a house. There must therefore be some cause higher than all other by virtue of which they all cause being and whose proper cause is being: and this cause is God. Now the proper effect of any cause proceeds therefrom in likeness to its nature... For this reason it is stated in *De Causis* (prop. ix) that none but a divine intelligence gives being, and that being is the first of all effects, and that nothing was created before it" (*On the Power of God*, III q. 7, a. 1 body). *On the Power of God*, trans. by English Dominican Fathers (London: Burns, Oates, and Washbourne, 1932-34; This edition will be used for all English translation references to the *On the Power of God*).

Abstraction occurs in two ways, first by composition and division where one understands that one thing exists in another, or that one thing can be separated from another. The second way of abstraction is simply by understanding one thing without considering other things that might be associated with it.[375] The latter is the way that the intellect abstracts the universal from particulars, or the intelligible species from phantasm; by considering the nature of the species apart from its individual qualities represented by the phantasms. The intellect simply does not consider the specific material conditions, and selects only those things that are common and that are of the very nature of what a thing is as given by its essential definition (comprehension) while leaving out any specific material details such as these blue eyes, this hand, etc. It is in the second operation of the intellect that composition and division takes place, as well as in the discursive power, which judges sensible intentions.

Based upon the theory of abstraction and incidental sensibles, taking Kant's understanding of Hume's position seriously, science itself is "saved" from Hume's critique in terms of universality and necessity since the causal relationship has a necessary connection to its given effect and this connection is itself universal since it can be common among many different effects. In agreement with Hume, common sensory impressions and repeated intentions form what Hume considered to be custom by which one perceives this effect and its efficient cause.

In disagreement with Hume, from these repeated and common sensory impressions, a universal and necessary *connexion* is apprehended

---

375 "Abstraction may occur in two ways: First, by way of composition and division; thus we may understand that one thing does not exist in some other, or that it is separate therefrom. Secondly, by way of simple and absolute consideration; thus we understand one thing without considering the other. Thus for the intellect to abstract one from another things which are not really abstract from one another, does, in the first mode of abstraction, imply falsehood… the things which belong to the species of a material thing, such as a stone, or a man, or a horse, can be thought of apart from the individualizing principles which do not belong to the notion of the species. This is what we mean by abstracting the universal from the particular, or the intelligible species from the phantasm; that is, by considering the nature of the species apart from its individual qualities represented by the phantasms…the mode of understanding, in one who understands, is not the same as the mode of a thing in existing: since the thing understood is immaterially in the one who understands, according to the mode of the intellect, and not materially, according to the mode of a material thing" (*ST* I, q. 85, a. 1, reply 1).

by the intellect as universal and as necessary through the incidental *connexion* made by the sensory cogitative power given by repeated incidental cause and effect *connexions* or *by a single incidental sensible connexion immediately apprehended as in perceiving Socrates move one perceives that Socrates is alive.* The *connexion* does in fact exist as one object touches another and thus can be seen, but the incidental causal association between the effect and its specific efficient cause is incidental to any such *connexion* or point of contact or contiguity between two objects in existence. One can see billiard ball A touch billiard ball B, but the cause of B's motion is incidental to the motion of B itself and the associated contact between A and B.

It is this incidental causal *connexion* that Hume failed to properly develop that led to Kant awaking from his Dogmatic slumber. Kant improperly expanded the function of understanding to include the cognitive function of the cogitative faculty. Hence for Kant the understanding itself would become responsible for empirical and *a priori* rules by which experience was to be understood rather than apprehended by a cognitive faculty. While a dichotomy between experience and pure reason remained in the transcendental philosophy inconsistent with the proper order of human knowing and experience itself, any distinction between the functions of the inferior apprehensive faculties such as memory, imagination, and the cognitive faculty and the superior apprehensive faculties of the intellect became non-existent in the Kantian noetic.

Again, the cogitative power or cognition in the act of perception prepares the sensible forms and incidental intentions for abstraction by the light of the intellect where the universal quiddity or whatness is assimilated to the intellect. In this, the possible intellect *becomes all things* and the agent active intellect *makes all things* in the act of assimilation. Klubertanz describes this cognitive process:

> One experience impressed into the memory by the
> cognizant sense, serves as a focal point around which other

similar experiences can gather… Sensory impressions pass through sense awareness with a bewildering rapidity… But in the course of all this confusion, there comes an experience which is recognized… At once the organism reacts to it as a whole, and that experience is impressed consequently upon the memory. Later that experience is repeated again, in a slightly different form, but with enough likeness to meet with a similar reaction under similar circumstances… memory comes into play, modifying, completing, and making firm the image received into the imagination. Once the three powers of imagination, memory, and the cognizant sense supervene upon the primary synthesis of the common sense, we can have the secondary, complete synthesis known as perception.[376] For the cognizant sense unifies the combined

376 Likewise for the synthetic operation of the phantasia see Dorothea Frede: "phantasia…first acts as a kind of synthesizer of individual sense-impressions, second it supplies the sensory underpinnings of thought" (see Dorothea Frede, "Aquinas on Phantasia" in *Ancient and Medieval Theories of Intentionality*, Dominik Perler (eds.), (Leiden: Brill, 2001), 155-183; 157). Frede discusses the role of phantasmata and sensory impressions in synthesis and concludes that the phantasia is "a synthesis of what I perceive right now and what I have perceived a second ago and so on. Kant describes very nicely how such a synthesis of a manifold takes place when he describes how we look up and down a house (*Critique of Pure Reason*, A190ff)."

Frede not only argues for a synthesizing role to be attributed to the phantasia in Aristotle, but also indicates that most of the functions of consciousness is to be attributed to the common sense: "Scholars have in recent years drawn attention to the integrative role of the 'common sense',…It seems that it fulfills most functions that we ascribe to consciousness." Frede mentions that the common sense is "responsible for the *koina aistheta*, the motion we both see and feel, or for the objects of accidental perception, as when we perceive the white thing as the son of Diares." (See Dorothea Frede, "The Cognitive Role of Phantasia in Aristotle" in *Essays on Aristotle's De Anima*, Martha C. Nussbaum and Amelie Oksenberg Rorty (eds.), (Oxford: Clarendon Press, 1997), 282-295). Although Alfred Leo White, Jr. agrees that the common sense has a role in discriminating between different proper senses and is aware of our perceptual operation, the common sense does not integrate or construct a synthesis of proper sensible qualities into one object. White indicates: "the common sense always performs these two operations together; that is, it senses an object while it perceives the very operations of the proper senses…through an intellectual operation we may judge that we are sensing, but prior to this kind of reflection we are already in some way aware of our perceptual operations. We have such an awareness at the level of the common sense" (Alfred Leo White, Jr. *The Experience of Individual Objects In Aquinas A Dissertation*, (Washington, D.C.: The Catholic University of America, 1997), 66). Later White argues that "Aquinas never says that the common sense perceives white and sweet as qualities of the same object: he simply refers to them as examples of proper sensible qualities that the common sense distinguishes from each other" (Ibid., 84). The important distinction is one of "perceiving different sensible qualities belonging to the same object, and perceiving them as belonging to the same object" (White, 85). White does consider the two operations of the common sense, sensing proper sensible qualities and perceiving the operations of the proper senses, to be actually one and the same operation not interwoven

data of the common sense, the imagination, and the memory, relating the object of sensation to its situation as a whole, as in the first instance, and to other experiences....

Thus, from sense through memory (and so by repetition) there arises perception, the *experimentum* of St. Thomas.... From the fully accomplished perception, which can even be called a 'quasi-universal' or 'implicitly universal' knowledge, because in a way it includes a number of singular experiences, there arises the strictly universal and abstract knowledge which is the act of the intellect. At the culmination of this process stands sense-perception (sometimes merely the elaborated phantasm), which, in subordination to the agent intellect (after the manner of an instrumental cause), produces the actual intelligible.[377]

---

but identical (White, 78-81). Further, most scholars agree that the *phantasia* is responsible for image formation and has a synthesizing role of the manifold illustrated by how one looks up and down a house as described by Kant. For Aquinas, however, it is not the common sense that discursively perceives the white thing as the son of Diares but rather it is cognition or the cogitative power that discursively perceives such incidental intentional objects.

White also explains, "an animal perceives its conjunction with an object, or it perceives its possession of that object. It seems that one perceives such a union with an object through the common sense's reflexive awareness... Feeling an object of touch involves not only a cognitive union with that object, but a real union as well; hence it seems that the common sense perceives this real union in perceiving the act and object of the sense of touch. That is, the common sense perceives that an object is really present to the perceiver's external sense organ" (White, 176). The concern here is not one of physiology, i.e., where such a common sense might be located in neuro-physiology, but instead the concern is one of function in cognitive psychology. Additional research is required in two areas: 1) how such features map to contemporary perceptual psychology, and 2) where this perceptual machinery is located within the brain itself and how specifically does it function according to contemporary neuro-cognitive psychology. It should also be noted that for Aquinas the functions of the common sense ascribed to consciousness is limited since it is the intellect that understands and is able to know that it understands in an act of "complete reflection," the common sense is only aware that it is sensing in an "incomplete reflection." (see *De Veritate* II, q. 19, a. 9; *De Veritate* I, a. 1, a. 9; *In Sent. Peter Lombard* II, d. 23, q. 1, a. 2, ad3; *ST* I q. 78, a. 4, ad2; *ST* I, a. 87, a. 3, ad3; *ST* I, q. 78, a. 4, ad2). The common sense power cannot reflect upon its own act, but instead it performs an incomplete reflection on the acts of the proper senses (White, 121). Since no sense power can reflect upon its own act, the sensitive power is not in and of itself consciousness even if it does fulfill certain functions that we ascribe to consciousness (White, 121; *SCG* IV, 11). Self-awareness is a reflexive activity of each higher power upon a lower power.

377 George P. Klubertanz, "The Internal Senses in the Process of Cognition," *The Modern Schoolman* XVIII no. 2 (1941), 30. Also, see Cornelio Fabrio, "Knowledge and Perception" *The New Scholasticism* XII (1938), 337-365; Dom Thomas and Verner Moore, "Gestalt Psychology and Scholastic Philosophy," *The New Scholasticism* VII (1933), 298-325m VIII (1934) 46-80.

Kant introduced the transcendental to account for universality and necessity (the *a priori*) and Kant introduced the synthetic notion to account for *a priori* knowledge of efficient causality in possible experience as conditioned by the transcendental philosophy. However, in Kant failing to realize that universality and necessity can be obtained by abstraction from sensibility in the case of proper and common *per se* sensibles and abstracting incidental sensibles from sensibility, Kant introduced unnecessarily the transcendental philosophy drawing from the Rationalist tradition. In introducing the notion of the *a priori* analytic, Kant failed to see that although universality and necessity of self-evident or analytic propositions are such due to intelligibility, one arrives at the *a priori* analytic again through abstraction from sensibility as discussed previously in the case of mathematical and logical concepts abstracted from the phantasm and formed in the imagination since at the level of natural knowledge there is nothing in the intellect that was not first in the senses.

Of course Kant was correct to suggest that one can have universality and necessity, and Kant was correct that there is synthetic and analytic aspects to knowledge. Therefore one can have *a priori* synthetic and analytic knowledge; however, this knowledge is not preconditioned by the transcendental philosophy nor is such knowledge entirely independent of experience. Rather this knowledge is obtained through the process of abstraction from sensibility following a similar representational model as proposed by Locke, following the common understanding of abstraction prior to early modern philosophy.

However, although Locke retained a notion of abstraction, Locke mistakenly concluded that secondary qualities or proper sensibles are merely perceptions or appearances or merely names rather than intrinsic qualities of or in a given object following late Medieval nominalism that had come to influence the Empiricist tradition of Galileo. Further, the early modern tradition failed to account for incidental sensibles. This turn in early modern philosophy led to the mistaken conclusion that all sensibility is entirely unreliable similar to Plato's mistaken view that sensation and perception is entirely unreliable due to contingent existence. Of course, this led Aristotle to propose that the proper sensibles and sensibility in general is much more reliable than Plato's account had suggested.

Interestingly, the move by both Plato and Kant was essentially the same for similar reasons. Both Plato and Kant accepted the contingent nature of sensibility and the unreliability of perception, and this forced both philosophers to propose an account that detached universality and necessity from contingent sensibility leading to a severe disparity or false dichotomy between sensibility and intelligibility. Aristotle and Aquinas held that a proper distinction between intelligibility and sensibility does in fact exist. However, Aristotle and Aquinas also maintained a uniformity of being between intelligibility and sensibility developing the hylomorphic view of substantial unity on one hand and Aquinas's later development of the Dionysian principle of hierarchy to account for the later Galenic and Arab developments in the medical sciences on the other hand. While retaining a distinction between sensibility and intelligibility, Aquinas like Aristotle before him was able to propose an alternative view to that of the materialist account that was unable to fully account for intelligibility purely in terms of sensibility as well as the Platonist view of his predecessors.

The rejection of a reliable means of apprehending intrinsic secondary qualities that were no longer considered anything more than mere qualities of perception, and also given Berkeley's perceptual account of primary qualities, Hume's critique using the critical method of the British Empiricists was inevitable. Likewise, the neglect of a proper account of incidental sensibles naturally led to Hume's critique of causality, substance and the like. Regardless of Hume's materialist leanings as a British Empiricist, Hume simply came to the inevitable and honest conclusion that from sensibility one cannot affirm the causal maxim and this naturally followed from the developments of early modern philosophy. With Berkeley and finally with Hume, nothing was left to modern philosophy except a turn to the ego, self-consciousness, and skepticism when it came to sensation and perception.

It is apparent, however, that one can conclude with Kant that a synthetic *a priori* knowledge of efficient causality is possible but in opposition to Kant such knowledge is possible when one obtains universal and necessary causality through abstraction from singular efficient causal events that are found to occur always and everywhere (or, always or for the most part)[378] by reflecting upon singulars in the process of synthetic unity at the level of sensibility and in the second operation of the intellect. In opposition to Kant, one thus attains synthetic *a priori* knowledge independent of the transcendental. Independent of the transcendental philosophy in the sense of being independent of the rules for the imaginative synthesis of the manifold or simply independent of the rules or schemata some of which are empirical and others are *a priori*.

Once a universal and necessary causal relationship is determined to exist by an act of the intellect in forming a judgment that similar cause and effect events have a similar efficient causal *connexion*, a universal and necessary causal relationship is judged to exist by the intellect and can then be applied through the cogitative power to any singular efficient causal relationship in existence to discursively determine its necessity and universality at the level of judgment. In this way, one can have a synthetic *a priori* knowledge of efficient causality independent of the transcendental.

Just as the intellect must return to the phantasms to know the singular, the intellect through the cogitative power must return to singular incidental efficient causal events to know the singular efficient causes and their related effects. By recalling similar efficient causal

---

378 Again universality is taken from something that occurs everywhere and always, but due to sensibility being attached to a given contingent context, circumstances may change and therefore it is reasonable to say that universality and necessity can be taken from something that occurs always or for the most part and this is not merely a statistical model for comparison. It is derived from the contingent nature of sensibility itself for Aristotle and Aquinas. If not hindered, necessity and universality occurs always and everywhere, but if the possibility exists that the necessity might be hindered then it occurs always or in most cases.

incidental contiguous *connexions* such as bat A striking ball B and bat C striking ball D and in both cases the bat having been struck breaks a window one judges that the incidental efficient cause of the bat striking the ball and hitting the window broke the window. Once a universal and necessary judgment is made where striking a ball breaks a window then the intellect in returning to the singular conditions of this or that bat hitting a given ball and breaking a particular window, the intellect can judge the proper efficient cause of why the window broke.

Such a judgment is *a priori* since it is universal and necessary that in general every ball that is struck by a bat that in turn hits a window will break the window unless otherwise hindered. This is the case because of the essential qualities or nature of the ball, bat, window, the impact of the bat upon the ball, and the impact of the ball upon the window along with any motion involved. The judgment is synthetic in that the effect is not defined in the cause but is an extension of the cause insofar as the window breaking is a consequence of the ball being hit by the bat and striking the window, and the judgment is associated with experience. The efficient causal proposition might be 'that bat struck the ball and the ball struck the window consequently the window is broken'. The predicate is not contained in the subject and therefore it is not a self-evident proposition, but the predicate is a consequence or an extension of the subject and hence a synthetic judgment exists where the terms are apprehended from experience. It has conditional necessity since a proper set of circumstances must obtain for the proposition to be true. Therefore the proposition is universal and necessary while being synthetic therefore it is synthetic *a priori*.

Hence, synthetic *a priori* knowledge of efficient causality is possible when the dichotomy between sensibility and intelligibility is removed while retaining proper distinctions between sensibility and intelligibility. A causal relationship is judged to exist in the second operation of the intellect while the cogitative power judges the intentions of sensible efficient causal relationships. By the intellect affirming that a sensible efficient causal relationship occurs always, or for the most part, in the process of returning to the sensible conditions of human knowledge, the intellect judges a necessary relationship between a given cause and its effect. In judging that every effect has a cause, the intellect judges a universal and necessary causal maxim from repeated incidental sensible conditions taken from custom and experience through the memorative power and reminiscence. In this way, there is one intellective operation moving discursively between intelligibility and sensibility by which synthetic *a priori* knowledge of efficient causality obtains.

Again early modern philosophy had difficulty determining the causal *connexion* because incidental sensibles do not directly affect the external senses. Socrates may be this or that color, but being Socrates is incidental to being of this or that particular color. Likewise, being a friend is incidental to being Socrates, but in perceiving the color white one incidentally apprehends that this white object is Socrates through a process of perceptive and cognitive association. In apprehending that this is Socrates, the cogitative power discursively judges that Socrates is a friend from short or long term memory using reminiscence or phantasms. The cogitative power thus affirms that Socrates is a friend in a discursive process of cognitive association at the level of sensibility through reminiscence. The intellect apprehends as universal that one is alive by the cogitative power perceiving that Socrates is in motion. Hence the universal can be apprehended from the singular incidental sensible.

Again it is the *per se* sensibles that directly affect the external senses and these *per se* sensible are those that the intellect apprehends to be associated with an incidental. This or that color for instance affects directly the external senses but the association of the son of Diares with a given color is incidental to this or that color. Likewise, this or that particular causal event or that given effect for instance affects directly the senses as Hume realized. However, *what Hume failed to understand is that one perceives repeated incidental causality when a billiard ball is struck by the hand of Socrates insofar as the eye sees the proper sensible from which the common sensible of movement and the connection between one or more objects are apprehended and associated with efficient causality by the cognitive faculties.* However, the association or *connexion* of this event being an effect of that particular cause is incidental to the events themselves and yet apprehended incidentally.

Hume failed to see this indirect or incidental apprehension of the causal *connexion* since there was no longer an incidental category from which to draw this causal relationship from sensibility much less any reliable means to perceive proper or common sensibles (or, secondary and primary qualities). Thus Hume was no longer able to justify an intellective apprehension of incidental sensibles via the cogitative or cognitive faculty. Hume was thus forced to suppose that only from experience and by custom can one infer from belief the causal maxim and various other relationships independent of reason. However, such universal and necessary relationships properly belong to the intellect in apprehending incidental sensibles from experience and memory. *Per accidens* the cogitative power apprehends that this man is Socrates, or this effect is related to this cause, or that this particular individual is a friend without directly affecting the external senses except united with *per se* proper or common sensibles. It is the apprehensive powers, not the external senses, that apprehends incidental sensibles in themselves "immediately and without hesitation" from that which is sensed in itself:

> *Per accidens* that is sensed which does not affect the sense inasmuch as it is a sense, nor as it is this sense, but as joined to those things which of themselves affect the sense, as "Socrates," and "the son of Diares," and "friend," and other like things. These things are known in the universal by the intellect; in the particular, they are known by the discursive power in men, and by the estimative in other animals. Such things the exterior sense is said to sense, even though only *per accidens*, when from that which is sensed in itself, the apprehensive power, whose task it is to know them in themselves, immediately, without hesitation or reasoning knows them; as we see that someone lives from the fact that he speaks.[379]

---

379 *In IV Sent.*, d. 49, q2, a2, ed. *In III Sent.*, d26, q1, a2, sol., ed. Pierre Mandonnet, O.P., and M. F. Moos, O.P., Paris: Lethielleux, 1929-1933, vol. VII, pt. 2, pp. 1201-02 in George P. Klubertanz, S.J.,

Likewise Aquinas explains how incidental sensibles are related to objects of the intellect's knowledge in the universal:

> An indirect object of sense is that which does not act on the sense, neither as sense nor as a particular sense, but is annexed to those things that act on sense directly: for instance Socrates; the son of Diares; <u>a friend and the like which are the direct object of the intellect's knowledge in the universal,</u> and in the <u>particular are the object of the cogitative power in man,</u> and of the estimative power in other animals. <u>The external sense is said to perceive things of this kind, although indirectly, when the apprehensive power (whose province it is to know directly this thing known), from that which is sensed directly,</u> apprehends them at once and without any doubt or discourse (thus we see that a person is alive from the fact that he speaks): otherwise the sense is not said to perceive it even indirectly.[380]

By being annexed to those things that directly act on sense, incidental sensibles are known in their universality such as the son of Diares, or a friend, or being alive and yet in the particular are also the objects of the cogitative power due to one intellective operation between the intellect and the cogitative faculty. The discursive or cognitive faculty or the cogitative power reasons from an effect to its cause in apprehending a given causal connexion of the singular. Once apprehended and abstracted the intellect forms universal and necessary synthetic propositions of efficient causality in judgment. Aquinas explains that the cogitative or discursive apprehension of particular intentions or incidental sensibles takes place according to the Dionysian principle of hierarchy that "every inferior nature in its highest element touches the lowest element of the superior nature, according as it participates something of the superior nature, although deficiently;..."[381]

The Discursive Power (Ohio: The Messenger Press, 1952).

380 *ST* III, q. 92, a. 2 body.

381 *In III Sent.*, d26, q1, a2, sol., ed. Pierre Mandonnet, O.P., and M. F. Moos, O.P., Paris: Lethielleux, 1929-1933, vol. III, pp. 816-17 in George P. Klubertanz, S.J., *The Discursive Power* (Ohio: The Messenger Press, 1952), 153.

Combining the Dionysian principle of hierarchy with Aristotle's hylomorphic principle, and yet maintaining a distinction between the sensible and intellectual powers, Aquinas continues, "in apprehension as in sensitive appetite there is to be found something in which the sensitive part touches reason… apprehending those intentions which do not fall under sense, like friendship, hatred, and the like, *this belongs to the sensitive part according as it touches reason.* And so that part in men, in whom it is more perfect on account of its being joined to the rational soul, is called 'particular reason,' because it compares particular intentions; but in other animals, because it does not compare, but has the power of apprehending such intentions from natural instinct, it is not called "reason," but "estimation."[382]

The cogitative power compares particular intentions or particular sensible forms, and is therefore a sense power; however, the cogitative power considers these particular forms or intentions by discursive and comparative analysis which are function of the intellect.[383] Aquinas adapts the cogitative development of Averroës to Avicenna's notion of intentions. However, Aquinas makes both the agent and possible intellect to be in the composite individual and not a separate substance independent of the individual. Against Averroës Aquinas argues that the possible intellect is not shared in common by all individuals, and against Avicenna, Aquinas argued that the agent intellect does not infuse knowledge as a separate substance.

Aquinas separates the passive or cogitative faculty from the possible intellect indicating that the intellect as such cannot be mixed with a bodily organ since the intellect is incorporeal. Since nothing moves except by that which is in act, the agent intellect is that which moves the intellect to apprehend universals just as by analogy the sensible objects move the external senses to their proper perceptive acts of sensation. Averroës had maintained that the cogitative power is

---

382 Ibid., 153.
383 George P. Klubertanz, S.J., *The Discursive Power* (Ohio: The Messenger Press, 1952), 163.

that which distinguishes humans from the other animals and that the possible intellect is a separate substance shared in common with all humans, but against Averroës, Aquinas argued that it is the possible intellect as intellect that distinguishes humans from other animals and the possible intellect is a separate substance in the sense of being unmixed with a corporeal organ but not independent of the individual and of phantasms in particular.[384]

Aquinas in agreement with Averroës argued that the cogitative power or the passive intellect is itself corporeal since their needed to be a way to bridge between the corporeal internal senses and the incorporeal intellect. To make this philosophically possible, Aquinas also retained the Arab notion of "in contact with" (*continuatur*) as being consistent in some respect with the Dionysian principle of hierarchy, but Aquinas rejected the Arab and Latin's neo-platonist sense of emanation of forms by the agent intellect. Aquinas reformulated the Arab notion and argued that the contact is one between the incorporeal universal reason and the corporeal particular reason connected as one substantial composite subject. Aquinas explains this contact in terms of "touch" or "contact" or "overflow." We thus find in this analysis, "a sense power, just at the point of juncture of sense and intellect, which is called '*vis cogitativa*.'"[385] Aquinas writes:

> That power, which is called "*cogitativa*" by the philosophers, is on the boundary of the sensitive and intellective parts, where the sensitive part touches the intellective. For it has something from the sensitive part, namely, that it considers particular forms; and it has something from the intellective, namely, that it compares; and so it is in men alone. And because the sensitive part is better known than the intellective, for this reason, just as the determination of the

384 Edward P. Mahoney, "Aquinas's Critique of Averroës' Doctrine of the Unity of the Intellect," in *Thomas Aquinas and His Legacy*, ed. David M. Gallagher (Washington D.C: The Catholic University of America Press, 1994), 83-105. Also, see *Saint Thomas Aquinas On the Unity of The Intellect Against the Averroists*, trans. and intro. by Beatrice H. Zedler (Milwaukee: Marquette University Press, 1968).
385 George P. Klubertanz, S.J., *The Discursive Power* (Ohio: The Messenger Press, 1952), 163.

intellective part is denominated from the sense, as was said, so every comparison of the intellect is named from "*cogitatio*."[386]

Aquinas describes a triple order between the various powers of the soul. The <u>first</u> <u>order</u> is between the intellectual and sensitive powers whereby the intellectual powers direct and command the sensitive powers.[387] The <u>second order</u> is that of nutrition and generation, whereby the sensitive powers prepare phantasms and sensible forms for abstraction of the intelligible species by the light of the agent intellect.[388] The <u>third order</u> is where the various powers interact in their operations.[389] Thus the relationship is one of preparation on the side of the internal senses consistent with the second order and one of instrumental causality where the intellect directs and commands the sensitive powers consistent with the first order. There is also interaction between the various cognitive faculties where certain faculties may be of a higher order than other faculties and where the action of one power is caused by the action of another as in the case of the action of the imagination being caused by the action of the senses, and this of course is consistent with the third order where various faculties interact.[390]

Therefore Aquinas's development of the hylomorphic view of the mind-body relationship, the Dionysian principle of hierarchy, and the reformulation of the notion of *continuatio* suggest a composite unity where there are no gaps between sensibility and intelligibility in response to Brentano's critique. Intentionality, the process of abstraction, the relation between universals and particulars, sensible and intelligible forms and intentions, and the cognitive psychology of Aquinas have been discussed in detail in the previous chapter.

---

386 *In III Sent.*, d23, q2, a2, q1, ad3 sol., ed. Pierre Mandonnet, O.P., and M. F. Moos, O.P., Paris: Lethielleux, 1929-1933, vol. III, p. 727 in George P. Klubertanz, S.J., *The Discursive Power* (Ohio: The Messenger Press, 1952), 163.
387 *ST* I q. 77, a. 4, body.
388 Ibid.
389 Ibid.
390 *ST* I q. 77, a. 7.

This chapter concluded that *a priori* synthetic knowledge of efficient causality is possible using cognitive psychology and a moderate realist account where an incidental causal *connexion* can be apprehended from sensibility through the cogitative power or cognition in apprehending and judging sensible intentions and these intentions in turn form a universal and necessary knowledge of efficient causality. A chapter not included in this thesis likewise develops the universal and necessary relationship between modal propositions and experience thereby developing a cogent argument for the synthetic *a priori* nature of modal propositions.

It has been cogently argued that one can have synthetic *a priori* knowledge of efficient causality and necessity by virtue of incidental sensibles, independent of the transcendental philosophy. Likewise a natural knowledge of God begins from sense according to Aquinas,[391] and this is independent of the transcendental. Independent of the transcendental philosophy in the sense of being independent of the rules for the imaginative synthesis of the manifold or simply independent of the rules or schemata some of which are empirical and others are *a priori*. The following chapter will argue that one can know from a synthetic *a priori* knowledge of efficient causality that God exists and know those things that *necessarily* belong to God. It will be argued that a natural knowledge of God's existence from synthetic a priori knowledge of efficient causality is in fact possible in opposition to the accounts given by Kant and Hume.

---

391 *ST* I q. 12, a. 12, body.

Chapter 6

SYNTHETIC A PRIORI KNOWLEDGE OF GOD'S EXISTENCE FROM EFFICIENT CAUSALITY AND NECESSITY

The Synthetic a priori and God's Existence

It has been cogently argued how it is possible to have synthetic *a priori* knowledge of efficient causality independent of the Kantian transcendental explaining that the *intentio non-sensata* and the mental act of the *vis cogitativa* make possible an awareness of individual incidental sensibles of efficient causality and necessity that go beyond inner sense as the cognitive act participates in a single intellectual operation. In demonstrating a synthetic *a priori* knowledge for the existence of God independent of the Kantian transcendental, motion taken in the broad and restricted sense will be used in Aquinas's arguments from motion, efficient causality, and from natural necessity. In this way, motion from potentiality to actuality implies agencies and will be the point of departure for the second and third ways of Aquinas in arguing for the existence of God.

Likewise it has been cogently argued that the Kantian transcendental was simply an unfortunate and unnecessary development of early modern philosophy. Against the transcendental metaphysics and deduction it has been argued from Aquinas that the categories are abstracted immediately and continually from sensibles and are *a priori* to the extent that the *predicaments* as 'categories' are universal and necessary having been abstracted from the sensible conditions of human experience. The categories structure conscious reality to the extent they are abstracted from existence and are categories of being.

Efficient causality and any related necessity is apprehended by the cogitative power as the *per se* sensibles act directly upon the senses and the *per accidens* sensibles act upon the powers of perception including imagination, memory, and the intellective cogitative power in perceiving the incidental sensible relationships of efficient causality and natural necessity and possibility in natural things. The cogitative power prepares the incidental sensible forms of efficient causality for abstraction as a universal causal relationship and a given *connexion* between a given cause and its effect by perceiving the congruity between a given cause and its effect. Once apprehended, the intellect places the effect as the middle term in place of the essence of sensible things in arguing for God's existence from efficient causality.[392]

Aquinas accepted that Aristotle's categories divide real being, but also maintained that being is not divided univocally into the ten predicaments or categories. Instead being [*ens*] is divided according to the diverse modes of being [*essendi*] and the modes of being [*essendi*] are proportional to the modes of predication and consequently the ten genera or kinds of being [*entis*] are called the ten predicaments

---

392 Aquinas explains: "Reply OBJ 1: The existence of God and other like truths about God, which can be known by natural reason, are not articles of faith, but are preambles to the articles; for faith presupposes natural knowledge, even as grace presupposes nature, and perfection supposes something that can be perfected. Nevertheless, there is nothing to prevent a man, who cannot grasp a proof, accepting, as a matter of faith, something which in itself is capable of being scientifically known and demonstrated.

Reply OBJ 2: When the existence of a cause is demonstrated from an effect, this effect takes the place of the definition of the cause in proof of the cause's existence. This is especially the case in regard to God, because, in order to prove the existence of anything, it is necessary to accept as a middle term the meaning of the word, and not its essence, for the question of its essence follows on the question of its existence. Now the names given to God are derived from His effects; consequently, in demonstrating the existence of God from His effects, we may take for the middle term the meaning of the word "God".

Reply OBJ 3: From effects not proportionate to the cause no perfect knowledge of that cause can be obtained. Yet from every effect the existence of the cause can be clearly demonstrated, and so we can demonstrate the existence of God from His effects; though from them we cannot perfectly know God as He is in His essence" (*ST* I, q. 2, a. 2 Rp 1, 2, 3).

or categories.[393] Predicaments constitute categories of terms that can be predicated of other terms and whenever something is predicated of something it falls under one of the ten categories. For when we predicate something of another, we say that this existing thing is predicated of that existing thing, and hence the ten genera or kinds of being [*entis*] are called the ten predicaments or categories. Aristotle makes clear that in nature, motion exists always in respect to substance, quantity, quality, or place or any of the other *predicaments* since there is nothing over and above the *predicaments* in nature.[394]

Insofar as a thing is denominated by the intelligibility [*ratio*] of an agent cause, there is the predicament of passion and action. The intelligibility [*ratio*] of motion or efficient causality occurs not only by perceiving motion in the nature of external things, but also by that which the intellect [*ratio*] apprehends through incidental perception. In nature, motion is an imperfect act moving from potency to act and

393 Aquinas explains in discussing the problem of whether action and passion make up one or two categories explains: "Being [*ens*] is not divided univocally into the ten predicaments as genera are divided into species. Rather it is divided according to the diverse modes of existing [*essendi*]. But modes of existing are proportional to the modes of predicating. For when we predicate something of another, we say this is that. Hence the ten genera [kinds] of being [*entis*] are called the ten predicaments [categories]. Now every predication is made in one of three ways. One way is when that which pertains to the essence is predicated of some subject, as when I say Socrates is a man, or man is animal. The predicament of substance is taken in this way.

Another mode is that in which that which is not of the essence of a thing, but which inheres in it, is predicated of a thing. This is found either on the part of the matter of the subject, and thus is the predicament of quantity (for quantity properly follows upon matter--thus Plato also held the great to be on the part of matter), or else it follows upon the form, and thus is the predicament of quality (hence also qualities are founded upon quantity as colour is in a surface, and figure is in lines or in surfaces), or else it is found in respect to another, and thus is the predicament of relation (for when I say a man is a father, nothing absolute is predicated of man, but a relation which is in him to something extrinsic).

The third mode of predication is had when something extrinsic is predicated of a thing by means of some denomination. For extrinsic accidents are also predicated of substances, nevertheless we do not say that man is whiteness, but that man is white. However, to be denominated by something extrinsic is found in a common way in all things, and in a special way in those things which pertain only to man" (emphasis added; *In Phys.* III, Lect. 5, Sect. 322).

394 Aristotle explains: "There is no such thing as motion over and above the things. It is always with respect to substance or to quantity or to quality or to place that what changes changes. But it is impossible, as we assert, to find anything common to these which is neither 'this' nor quantity nor quality nor any of the other predicates. Hence neither will motion and change have reference to something over and above the things mentioned; for there is nothing over and above them" (*Ari: Physics* III, 200b33-201a3; all citations of *Physics* are from *The Complete Works of Aristotle The Revised Oxford Translation* Vol I, Physics, ed. by Jonathan Barnes and rev. by R. P. Hardie and R. K. Gaye (New Jersey: Princeton University Press, 1984) unless otherwise noted).

is a certain beginning of the perfect act in that which is moved or the effect. For example, in something being whitened, something white has already begun to be white. Aquinas maintained that efficient causality is the principle which acts, or that whence the principle of motion is, and that for an imperfect act to have the *ratio* of motion, it is necessary for the intellect to understand that efficient causality or the principle of motion is a mean between two extremes.[395]

395 Aquinas explains: "To understand this we must consider that the efficient cause, which acts by motion, of necessity precedes its effect in time; because the effect is only in the end of the action, and every agent must be the principle of action" (*ST* I, q. 46, a. 2 Reply 1). Aquinas explains that an efficient cause must necessarily precede an effect in time since the effect is only in the end of the action of the efficient cause, and the efficient cause as an agent must be the principle of action and the efficient cause acts by motion. Again: "We must therefore say otherwise, that an efficient cause is twofold, principal and instrumental. The principal cause works by the power of its form, to which form the effect is likened; just as fire by its own heat makes something hot. ... But the instrumental cause works not by the power of its form, but only by the motion whereby it is moved by the principal agent: so that the effect is not likened to the instrument but to the principal agent: for instance, the couch is not like the axe, but like the art which is in the craftsman's mind. ..." (*ST* III, q. 62, a. 1, body par. 3/3). In this case a principle cause works by the power of its form, and the effect is like the form of the efficient cause. It is by the form of heat that fire moves something to become hot. However, in the case of an instrumental cause, the instrumental cause works by motion being moved by a principle agent.

Similarly, Avicenna argues that an efficient cause of motion from potency to act is that which enuces a form from the potentiality of matter by means of motion. The cutler being the efficient cause of a knife enduces the form of the knife from the potentiality of matter by means of producing the knife as an artifact or cutting instrument. If an effect depends upon its cause for its existence, if the cause is removed the motion that sustains the existence of the effect will likewise be removed. Avicenna explains: "As Avicenna says, no effect can remain if its proper cause is removed. Now, certain inferior causes are causes of becoming; others are causes of existing. A cause of becoming is that which educes a form from the potentiality of matter by means of motion, such as a cutler who is the efficient cause of a knife. A cause of a thing's existing, however, is that upon which the act of existence of a thing essentially depends, as the existence of light in the air depends upon the sun. Now, if the cutler is removed, the becoming of the knife ceases, but not its existence. However, if the sun is taken away, there ceases the very existence of the light in the air. Similarly, if God's action ceases, the existence of a creature utterly ceases, since God is the cause not only of a thing's becoming but also of its existence" (*De Veritate* I, q. 5, a. 8, reply 8).

Again the clear relationship between motion and efficient causality is given, "From this it is plain, therefore, that there are three principles of nature; matter, form and privation. But these are not sufficient for generation. What is in potency cannot reduce itself to act; for example, the bronze which is in potency to being a statue cannot cause itself to be a statue, rather it needs an agent in order that the form of the statue might pass from potency to act. Neither can the form draw itself from potency to act. I mean the form of the thing generated which we say is the term of generation, because the form exists only in that which has been made to be. However, what is made is in the state of becoming as long as the thing is coming to be. Therefore it is necessary that besides the matter and form there be some principle which acts. This is called the efficient, moving or agent cause, or that whence the principle of motion is. Also, because, as Aristotle says in the second book of the Metaphysics, everything which acts acts only by intending something, it is necessary that there be some fourth thing, namely, that which is intended by the agent; and this is called the end" (*De principiis naturae ad fratrem Sylvestrum*, trans. by R. Kocourek (St. Paul: North Central, 1948), Sct. 18, p. 13; emphasis added). Hence, the principle that acts is the efficient or moving or agent cause. In other words, the principle that acts is the principle of

On one end is potency to act and on the other end is act to potency. For Aristotle, motion is the fulfillment of what is potential *qua* potential between the actuality of an agent and the actuality of the patient.[396] In reason apprehending motion as being a mean between two termini, the intelligibility [*ratio*] of cause and effect is already implied since something cannot be reduced from potency to act except by some agent cause. Hence, the two predicaments, action and passion, are taken from the intelligibility [*ratio*] of the agent cause and its effect.[397]

Both Hume and Kant failed to see this connection between sensibility and intelligibility that had been lost in the early modern period. According to Kant "The question was not whether the concept of cause was right, useful, and even indispensable for our knowledge of nature, for this Hume had never doubted; but whether that concept could be thought by reason *a priori*, and consequently whether it

---

motion considered to be the efficient, or moving, or agent cause and this is the case because something in potency cannot reduce itself to act.

396 Aristotle indicates: "it is the fulfillment of what is potential as potential that is motion" (*Arist. Phys.* III, 201b7-201b15).

397 Aristotle had defined nature as the principle of motion and change, motion as the fulfillment of what is potentially the case. Motion Aristotle defined as the fulfillment of what is potentially the case: "Nature is a principle of motion and change...the fulfillment of what is potentially, as such, is motion-- e.g. the fulfillment of what is alterable, as alterable, is alteration; of what is increasable and its opposite, decreasable (there is no common name for both), increase and decrease; of what can come to be and pass away, coming to be and passing away; of what can be carried along, locomotion" (Ari: *Physics* III, 200b11, 201a10-201a14). However, Aquinas explains the intelligibility of motion as follows: "For the intelligibility [*ratio*] of motion is completed not only by that which pertains to motion in the nature of things, but also by that which reason [*ratio*] apprehends. For in the nature of things motion is nothing other than an imperfect act which is a certain incipience of perfect act in that which is moved. Thus in that which is being whitened, something of whiteness already has begun to be. But in order for the imperfect act to have the nature [*ratio*] of motion, it is further required that we understand it as a mean between two extremes. The preceding condition is compared to it as potency to act, and thus motion is called an act. The consequent condition is compared to it as perfect to imperfect or as act to potency. And because of this motion is called the act of that which exists in potency, as was said above. Hence, if an imperfect thing be taken as not tending toward something perfect, it is called a terminus of motion and will not be a motion in respect to which something is moved. For example, a thing begins to be whitened and the alteration is suddenly interrupted.

With reference to that which pertains to motion in the nature of things, motion is placed by reduction in that genus which terminates the motion, as the imperfect is reduced to the perfect, as was said above. But with reference to that which reason [*ratio*] apprehends regarding motion, namely, that it is a mean between two termini, the intelligibility [*ratio*] of cause and effect is already implied. For a thing is not reduced from potency to act except by some agent cause. And in respect to this motion belongs to the predicament of action and passion. For these two predicaments are taken in respect to the intelligibility [*ratio*] of agent cause and effect, as was said above" (*In Phys.* III, Lect. 5, Sect. 324).

possessed an inner truth, independent of experience, implying a perhaps more extended use not restricted merely to objects of experience" (*Prolegomena*, 6-7). Even if Hume was looking for an extended use of causality not restricted merely to objects of experience and that could only be thought by reason *a priori*, Hume rejected the possibility of an *a priori* notion of causality, universality, and necessity since these are arrived at from experience and custom independent of reason in the Humean Empiricist notion of causality.

Hume reasoned that one cannot by reason attain knowledge of efficient causality nor have any *a priori connexion* between a given cause and its effect. Neither Hume nor Kant were able to see, following the developments of the early modern period, that the *a priori* does not in fact need to be independent of experience, and the *a priori* is in fact apprehended from experience in both its universality and necessity.[398]

---

398 This position may be in opposition to the account given by Kripke, Casullo, and other analytic philosophers who like Hume and Kant see experience to be independent of the *a priori*. Instead analytic philosophers focus upon the relation of necessity and the *a posteriori* and discuss how modal *a posteriori* propositions can be either necessary propositions or necessarily true propositions with the emphasis being on the truth value of the latter type of modal propositions. This is the case since modal propositions such as 'the table is made of ice' is necessarily true if the table is in fact made of ice and such inductive proposition suggest some form of modal necessity. Logical beings of reason or second intentions can be *a priori* in a qualified Kantian sense just as imaginary beings of reason, but even these are not entirely independent of experience nor are they conditioned by the categories of being entirely independent of experience. For example, I can reason that 'all black dogs are black' but without having had apprehended from experience the categorical accidental quality of black and the substance 'black dog', the proposition itself would be meaningless. Likewise, contrary to Kant, *a priori* synthetic propositions are apprehended by abstracting universality and necessity from experience for such propositions are immediately associated with experience and this is consistent with the natural mode of human knowledge suggesting that the transcendental is not required. Modern Analytics often fail to realize that for Kant and Hume there is no connection between reason and experience in the *a posteriori* since the *a posteriori* is independent of reason in the empiricism of Hume just as the *a priori* is independent of experience in the rationalism of Kant. It is this false dichotomy that is the legacy of the early modern period. Therefore for neither Kant nor Hume can there be *a posteriori* necessary modal propositions nor *a posteriori* necessary truth values. Likewise, apart from the synthetic *a priori* there can be no *a priori* knowledge associated with possible experience in Kant. Therefore the categories of *a priori* and *a posteriori* naturally raise the question, are such notions in philosophical discourse even defensible. A discussion of the contemporary understanding of the *a priori* and *a posteriori* as understood by Albert Casullo, Kripke, and analytic philosophy in general goes beyond the scope of this thesis, but it is clear that Kripke and Casullo have begun to see various "ambiguities" associated with the *a priori* of the early modern turn as analytic philosophy attempts to use the *a priori* notions of Kant in formal logic following Frege and Russell. One should see the excellent work by Albert Casullo, *A Priori Justification* (Oxford: Oxford University Press, 2003), 185-209; Kripke, *Naming and Necessity*, 34ff, 104ff; Kripke, "Identity and Necessity," 152ff.

From this epistemological reformulation, a natural knowledge of God from efficient causality and necessity will be argued for from Aquinas's second and third way. [399]

The Medieval Triad and the Limits of Reason

Aquinas makes clear that what is first known to the intellect are the categories of being and particularly natural substance or what a thing is and not *immaterial created substance*. [400] In answering the

---

399 Aquinas indicates that Damascene had said "that God's existence is the same as his essence, and as his essence is unknown so also is his existence" (*On the Power of God*, III q. 7, a. 1 reply 1), but Aquinas indicates that this should be taken in relation to the truth of a proposition. Aquinas argues that in point of fact "we know that God is, because we conceive this proposition in our mind from his effects" and it is in this way that the proposition "God is" reflects the truth of a proposition. One may also say that "blindness is" because it is true that a man is blind even though blindness is a privation of sight (*On the Power of God*, III q. 7, a. 1 reply 1). The truth of the proposition that God exists is in the intellect as one apprehends God's existence from his effects and finds a proper correspondence between a given effect and its cause. In this way, a natural knowledge of God is possible according to man's proper mode of knowing which is from sensible things against the Kantian transcendental.

Further, a proposition is self-evident or analytic when a predicate is defined in the subject. However, Aquinas again makes the distinction: "A proposition may be self-evident in itself and yet not self-evident to this or that individual; when, to wit, the predicate belongs to the definition of the subject, which definition is unknown to him: thus if he knew not what is a whole, he would not know this proposition, A whole is greater than its part. The reason is that such propositions become known when their terms are known (*Poster. Anal.* i). Now this proposition, God is, is in itself self-evident, since the same idea is expressed in both subject and predicate: but with regard to us it is not self-evident, because we know not what God is: so that for us it needs to be proved, ..." (*On the Power of God*, III q. 7, a. 1 reply 11). Even though a given proposition may be in fact self-evident or analytic being both necessary and universal in itself, due to the predicate being defined in the subject, the proposition may not in fact be self-evident or necessary to us if we do not know the terms of the proposition. Hence, God's existence may in fact be in itself a self-evident necessary analytic proposition; however, it may not be self-evident to us that this is the case. Therefore Aquinas rejects the idea that God's existence is self-evident to us even though it is self-evident in itself, and thus in the order of reason it must necessarily be demonstrated from his effects.

400 Aquinas explains: "It is evident too that substance is first in the order of knowing, for that is first in the order of knowing which is better known and explains a thing better. Now each thing is better known when its substance is known rather than when its quality or quantity is known; for we think we know each thing best when we know what man is or what fire is, rather than when we know of what sort it is or how much it is or where it is or when we know it according to any of the other categories. For this reason too we think that we know each of the things contained in the categories of accidents when we know what each is; for example, when we know what being this sort of thing is, we know quality; and when we know what being how much is, we know quantity. For just as the other categories have being only insofar as they inhere in a substance, in a similar way they can be known only insofar as they share to some extent in the mode according to which substance is known, and this is to know the whatness of a thing" (*In Meta.* II, 7.1259). Also, *In Meta.* VII, 4.583, 1336.

Aquinas explains in the foreword of his *Commentary on the Posterior Analytics of Aristotle* that there are three operations of the intellect and the first operation of the intellect is concerned with apprehending the categories or predicaments: "Now there are three acts of the reason, the first two of which belong

question whether God is the first object known by the human mind Aquinas explains:

> On the contrary, "No man hath seen God at any time" (Jn. 1:18). I answer that, Since the human intellect in the present state of life cannot understand even immaterial created substances (ST q. 84, a. 1), much less can it understand the essence of the uncreated substance. Hence it must be said simply that God is not the first object of our knowledge. Rather do we know God through creatures, according to the Apostle (Rm. 1:20), "the invisible things of God are clearly seen, being understood by the things that are made": while the first object of our knowledge in this life is the "quiddity of a material thing," which is the proper object of our intellect, as appears above in many passages (*ST* I, q. 84, a. 7; *ST* I, q. 85, a. 8; *ST* I, q. 87, a. 2, ad 2; emphasis added).[401]

As Buersmeyer indicates, Aquinas accepted the medieval triad of the *modi essendi, modi intelligendi* and the *modi significandi* with the source being Aristotle's *Peri Hermeneias* where words are conventional signs of concepts, and concepts are the likeness of the forms of things.[402] The *modi significandi* provided a means to talk about God due to the limitations of the human intellect.[403] Aquinas's metaphysics of predication suggests that the *modi significandi* of concrete names signify *per modum completi partipantis* and abstract names signify *per modum diminuti et partis formalis.*[404]

to reason regarded as an intellect. One action of the intellect is the understanding of indivisible or uncomplex things, and according to this action it conceives what a thing is. And this operation is called by some the informing of the intellect, or representing by means of the intellect. To this operation of the reason is ordained the doctrine which Aristotle hands down in the book of *Predicaments*, [i.e., Categories]. The second operation of the intellect is its act of combining or dividing, in which the true or the false are for the first time present. And this act of reason is the subject of the doctrine which Aristotle hands down in the book entitled *On Interpretation*. But the third act of the reason is concerned with that which is peculiar to reason, namely, to advance from one thing to another in such a way that through that which is known a man comes to a knowledge of the unknown. And this act is considered in the remaining books of logic" (*Foreword in the Commentary on the Posterior Analytics of Aristotle, trans. by F. R. Larcher (Albany, NY: Magi Books, 1970)*).

401 *ST* I q88, a3, body.
402 Keith Buersmeyer, "Aquinas on the 'Modi Significandi'" *The Modern Schoolman* (March 1987), 73.
403 Ibid., 83.
404 Ibid., 81-82; *In Sent.* III, d. 7, q. 1, a. 1 and *Topics* II, 109a, 34ss.

In the *De Enti et Essentia* Aquinas maintains that the *modus significandi* of concrete names include the metaphysics of participation.[405] Buersmeyer explains for example: "when *homo* functions as a logical subject, it supposes or stands for a concrete individual participating in the substantial form of humanity. When *homo* functions as a logical predicate, it signifies "*habens humanitatem indistincte*" and can be predicated of any individual which participates in the act of being a man."[406]

However, this is not a one to one correspondence between the concrete name (*dictio*) and the external object (*res*), but rather a reflection of the participation of creatures. For example, "*Socrates est homo*" is a true predication because Socrates participates in the form of humanity, and this in turn limits the act of being human to his concrete individuality.[407] Although this is not a naïve one to one correspondence between *dictio* and *res*, there is a conformity between the *modi significandi* of concrete names and the *modi essendi* of individuals *participating in* or *having* a given nature.[408]

Aquinas explains, "There is found in any creature a distinction between 'the individual having' and 'what is had'… the individual 'has' its nature, as a man 'has' humanity and as he 'had' in addition the act of *esse*."[409] Buersmeyer explains, "the concrete name '*homo*' signifies as 'having the nature of humanity,' and the verb '*est*' signifies that this nature is really 'had' or possessed by the individual, since it signifies the primary act of '*esse*' which places any form in act."[410] The concrete name can signify an accident or essential predication, but in the case of accidental predication, participation in act signifies having some accident and in losing that act, one continues to exist as an individual.[411]

However, in essential predication what is signified in the *modi significandi* is the *modi essendi* of an essential intrinsic participation in

---

405 Ibid., 81.
406 Ibid., 81-82.
407 Ibid., 82; See *Compen. Theol.*, ch. 9, n. 580.
408 Ibid., 82.
409 Ibid., 82. *In Periherm* I, lect. 5, n. 73.
410 Ibid., 82.
411 Ibid., 82.

or having a given act of existence or nature without which the subject cannot exist.[412] Socrates can lose the act of being pale and continue to exist, but Socrates cannot lose the act of being human and continue to exist *qua* human. Aquinas understands the imperfect relationship between *dicto* and *res* and thus Thomas holds that the *modi intelligendi* is a necessary intermediary between *modi significandi* and the *modi essendi*.

The *re/ratione* distinction in Aquinas is not the synthetic *a posteriori/*synthetic *a priori* transcendental epistemological distinction of Kant. In the case of Aquinas there are no Kantian transcendental categories that form the *ratio significata* nor are such categories projected upon existence since for Aquinas any such categories are apprehended by the intellect from existence since the intellects proper mode of knowing is through the senses from which the first thing that the intellect conceives is being [*ens*] and the proper object of the intellect is what a thing is.[413]

Kant understood that universals are supplied by the intellect, but Kant no longer had a reliable basis to discuss adequately the relationship between reason and experience. Kant rejected the three essential axioms of the *via antiqua*: 1) the thing known is in the knower according to the mode of the knower,[414] 2) the proper mode of human knowing is mediated by the senses and the sensible powers of perception because the thing known is in the knower according to the mode of the knower,[415] and therefore 3) the proper object of the intellect are the quiddities of sensible things.[416]

---

412 Ibid., 82-83.

413 "In the first operation the first thing that the intellect conceives is being [*in prima quidem operatione est aliquod primum, quod cadit in conceptione intellectus, scilicet hoc quod dico ens*], and in this operation nothing else can be conceived unless being [*ens*] is understood [*intelligatur*]…" (*In Meta.* 1.4 Lsn 6 Sct 605; emphasis added). "The proper object of the intellect is what a thing is: wherefore about this the intellect is not deceived except accidentally; whereas it is deceived about composition and division; even as the senses are always true about their proper objects, but may be deceived about others" (*SCG* I, 58).

414 Aquinas explains: "For knowledge is regulated according as the thing known is in the knower. But the thing known is in the knower according to the mode of the knower. Hence the knowledge of every knower is ruled according to its own nature" (*ST* I, q. 12, a. 4 answ).

415 Aquinas indicates: "since our intellect's knowledge, according to the mode of the present life, originates from the senses: so that things which are not objects of sense cannot be comprehended by the human intellect, except in so far as knowledge of them is gathered from sensibles" (*SCG* I, 3).

416 Aquinas clarifies "First, it can be taken merely according to its relation to that from which it first received its name. We are said to understand, properly speaking, when we apprehend the quiddity of

In the case of God, what belongs to the *res significata*, the absolute simplicity of Pure Act, is distinguished clearly from the *conceptus* or *ratio nominis* which is the conceptual meaning grounded in the *modi intelligendi*. Hence a distinction between what is *secundum rationem* and the *secondum rem* exists, but there remains a correspondence between the *ratio* and *rem* as Aquinas argues is necessarily the case for any knowledge of God to be known as true. As indicated below, Aquinas argues that truth obtains in judgment in conformity between *secundum rationem* and *secondum rem*. The intellect is assimilated to a given object and finds in judgment a conformity between what is known of an external object by the knowing subject and the same external object existing independent of the knowing subject.

For Aquinas the metaphysics of participation allowed one to predicate the same names properly of both God and creatures by analogy.[417] Since God and creatures do not share in the same form and due to the radical difference in being between the Creator and creature, predicating the same name properly of God and creatures is to signify analogously. It is the *modi intelligendi* in an act of judgment that makes

things or when we understand those truths that are immediately known by the intellect, once it knows the quiddities of things. For example, first principles are immediately known when we know their terms, and for this reason intellect or understanding is called "a habit of principles." The proper object of the intellect, however, is the quiddity of a thing. Hence, just as the sensing of proper sensibles is always true, so the intellect is always true in knowing what a thing is, as is said in *The Soul* (Aristotle, *De anima*, III, 6 (430a 26; 430b 27)). By accident, however, falsity can occur in this knowing of quiddities, if the intellect falsely joins and separates. This happens in two ways: when it attributes the definition of one thing to another, as would happen were it to conceive that "mortal rational animal" were the definition of an ass; or when it joins together parts of definitions that cannot be joined, as would happen were it to conceive that "irrational, immortal animal" were the definition of an ass. For it is false to say that some irrational animal is immortal. So it is clear that a definition cannot be false except to the extent that it implies a false affirmation. (This twofold mode of falsity is touched upon in the *Metaphysics* (Aristotle, *Metaph.*, {D}, 29 (1024b 18 seq.)). Similarly, the intellect is not deceived in any way with respect to first principles. It is plain, then, that if intellect is taken in the first sense--according to that action from which it receives the name intellect--falsity is not in the intellect.

Man's proper activity, however, is to understand through the mediation of sense and imagination. For the activity by which he fixes on intellectual things alone, passing over all lower things, does not belong to man as man, but in so far as something divine exists in him, as is said in the *Ethics* (Aristotle, *Ethica Nicomachea*, X, 7 (1177b 26)). Moreover, the activity by means of which he grasps only sensible things apart from understanding and reasoning does not belong to him as man, but according to the nature which he has in common with the brute animals. Therefore, when man is transported out of his senses and sees things beyond sense, his natural mode of knowing is modified" (*De Veritate* II, q. 13, a. 1, body).

417 Keith Buersmeyer, "Aquinas on the 'Modi Significandi'" *The Modern Schoolman* (March 1987), 84.

possible the analogous predication between God and creature. Since the *res significata* of creatures fail to capture accurately the *res significata* of God and since there can be no direct recourse to God through the senses and imagination, one must resort to analogy or God's effects to speak about God. Since creatures have goodness or wisdom by participating in the transcendental perfections which are convertible with the Pure Act of *Esse Subsistens*, concrete names signifying these perfections can be properly predicated of God.[418] Hence God is *per essentiam* what the creature is only *per participationem*.[419]

For Kant God cannot be known by means of the senses and thus God is known only through faith and hence no natural knowledge of God is available in Kant's epistemology, but for Aquinas although it is true that what God is cannot be fully known directly from sensibles, something about God such as his existence and attributes can be known from his effects through the imperfect participation of creatures by grasping the relation of other things to Him as when we say that God is the first cause or the sovereign good. Likewise, what God is not can also be known through this same imperfect participation.[420]

Although Kant maintained that we can only perceive possible experience in the manifold of empirical intuitions, for Aquinas one can know *res* through abstraction and predication where the likeness of

---

418 Ibid., 85.

419 Ibid., 85.

420 Aquinas explains: "… in the present life we know God only from his effects, as can be shown from what has been said above (*Faith, Reason, and Theology* q. 1, a. 2). Therefore, through natural reason we can know about God only what we grasp of him from the relation his effects bear to him, for example, attributes that designate his causality and his transcendence over his effects, and that deny of him the imperfections of his effects" (*Faith, Reason, and Theology*, q. 1, a. 4 body). Aquinas indicates: "Accordingly in every term employed by us, there is imperfection as regards the mode of signification, and imperfection is unbecoming to God, although the thing signified is becoming to God in some eminent way: as instanced in the term goodness or the good: for goodness signifies by way of non-subsistence, and the good signifies by way of concretion. In this respect no term is becomingly applied to God, but only in respect of that which the term is employed to signify. Wherefore, as Dionysius teaches, such terms can be either affirmed or denied of God: affirmed, on account of the signification of the term; denied, on account of the mode of signification. Now the mode of supereminence in which the aforesaid perfections are found in God, cannot be expressed in terms employed by us, except either by negation, as when we say God is eternal or infinite, or by referring Him to other things, as when we say that He is the first cause or the sovereign good. For we are able to grasp, not what God is, but what He is not, and the relations of other things to Him, as explained above (see *SCG* I, 14)" (*SCG* I, 30).

a thing is assimilated *secundum modum intellectus* not *secundum modum rei*. Hence, truth is found when what is conceived by the intellect conforms to what something is even though what is conceived may imperfectly represent something as it is in itself. Aquinas maintains that the *res significata* of God completely transcends what the human intellect can apprehend, but the human intellect can at least apprehend from God's effects God's existence and something about God and what God is not. Hence the human intellect can "reach past the creature to posit a *res* in the infinite God."[421] Aquinas explains, "Dionysius discusses the Divine Names (*Div. Nom.* i, iii) as implying some causal relation in God; for we name God, as he says, from creatures, as a cause from its effects."[422]

Aquinas further argues that the effects can signify the divine essence, "not comprehensively but imperfectly."[423] Aquinas explains the *res/ratio* distinction and its relation to a natural knowledge of God when discussing how terms can be hindered from being synonymous either because of the things signified, or because of the various aspects

---

421 Greggory, Rocca, O.P., "Res Significata and Modus Significandi", *Thomist*, 55(1991), 181-182.

422 *ST* I, q. 5, a. 2 reply 1

423 Aquinas explains: "Since every agent acts inasmuch as it is actual and consequently produces its like, the form of the thing produced must in some manner be in the agent: in different ways, however... When, however, the effect is improportionate to the power of the cause, the form is not of the same kind in both maker and thing made, but is in the agent in a more eminent way. Because according as the form is in the agent, the latter has the power to produce the effect: so that if the whole power of the agent is not reflected in the thing made, it follows that the form is in the maker in a more eminent way than in the thing made. This is the case in all equivocal agents,... Since then our intellect takes its knowledge from creatures, it is informed with the likenesses of perfections observed in creatures, namely of wisdom, power, goodness and so forth. Wherefore just as creatures by their perfections are somewhat, albeit deficiently, like God, even so our intellect is informed with the species of these perfections. Now whenever an intellect is by its intelligible form assimilated to a thing, that which it conceives and affirms in accordance with that intelligible species is true of that thing to which it is assimilated by its species: inasmuch as knowledge is assimilation of the mind to the thing known. Hence it follows that whatsoever the intellect informed with the species of these perfections conceives or asserts about God, truly exists in God who corresponds to each one of these species inasmuch as they are all like him. Now if such an intelligible species of our intellect were equal to God in its likeness to him, our intellect would comprehend him, and the intellect's conception would be a perfect definition of God, just as a walking animal biped is a perfect definition of a man. However, this species does not perfectly reflect the divine essence, as stated above, and therefore although these terms which our intellect attributes to God from such conceptions signify the divine essence, they do not signify it perfectly as it exists in itself, but as it is conceived by us. Accordingly we conclude that each of these terms signifies the divine essence, not comprehensively but imperfectly" (*On the Power of God* III, q.7, a. 5 body). .

consequent to the conception of the intellect.[424] Aquinas maintains that these various aspects which are in the mind cannot be such that nothing corresponds to them on the part of the thing (res) since if there were nothing in God, either in himself or in his effect, that corresponded to these various aspects, the intellect would mistakenly attribute these aspects to God, and such propositions in judgment would be false, "which is inadmissible."[425] Aquinas maintains that this is inadmissible and therefore there must be a correspondence that allows true propositions to be made about God.

Aquinas admits that there are concepts in the intellect to which nothing correspond in existence such as genus or species, which are logical second intentions. However, Aquinas argues that God is not to be taken as a second intention concept. If one were to attribute to God some aspect not as he is but consequent of some intellectual conception independent of correspondence, when one says God is good, God would be understood as such, but God would not be so in reality and there would be no basis for asserting that God is such in reality.[426] Hence, one might form an opinion about what they understand God to be, but there would be no way to know anything about God as he actually exists.

---

424 Aquinas indicates "Now terms may be hindered from being synonymous either by reason of the things signified, or on the part of the notion conveyed by the term and to signify which the term is employed. Wherefore the terms which are applied to God cannot be hindered from being synonymous by reason of their signifying different things, according to what has been said above, but only by the various aspects consequent to the conception of the mind" (*On the Power of God* III, q.7, a. 6 body).

425 Aquinas maintains: "Now these various aspects which are in our mind cannot be such that nothing corresponds to them on the part of the thing: since the things which these aspects regard are ascribed to God by the mind. Wherefore if there were nothing in God, either in himself or in his effect, corresponding to these points of view, the intellect would be in error in attributing them to him, and all propositions expressive of such attributions would be false; which is inadmissible" (*On the Power of God* III, q.7, a. 6 body).

426 Aquinas argues: "Because just as the intellect understands things existing outside the mind, so does it, by reflecting on itself, understand that it understands them: wherefore just as the intellect has a conception or notion to which the thing as existing outside the mind corresponds, so has it a conception or notion to which the thing corresponds as understood: for instance, to the notion or conception of a man there corresponds the thing outside the mind, while nothing but the thing as understood corresponds to the notion or conception of the genus or species. But it is impossible that such be the meaning of these expressions that are applied to God: for in that case the intellect would not attribute them to him as he is in himself but as he is understood: and this is plainly false; for when we say God is good, the sense would be that we think him to be so, but that he is not so in reality" (*On the Power of God* III, q.7, a. 6 body).

Aquinas explains further that one should not attribute aspects of an effect to God as if to say that the effect determines what or who God is. For example, saying God is wise because he is the cause of wisdom is mistaken since it is the case that because God is wise that God therefore causes wisdom.[427] Likewise, one is not to attribute *actual causality* to God in saying God is and is the cause of goodness and therefore God is good. If this were the case, one could equally predicate to God the names of all the divine effects. For example, because God is and is the cause of heaven, God is heaven since he is the cause of heaven.[428] Again this is absurd and therefore is an improper correspondence between the terms of a given cause and its effect in the case of God.

Instead, Aquinas argues for *virtual causality*, where God is called good because he is and has the power to infuse goodness and hence the term good signifies that power. In this case, the power is a *supereminent likeness of its effect* even as the power of any equivocal agent. In conceiving goodness, the intellect conceives the likeness of goodness that is in God, and in conceiving that goodness, one conceives God because God is that goodness. Hence, something that is in God and that is God corresponds to the notion or conception of goodness.[429] The concept that the intellect has of a thing is not true unless that thing corresponds to that concept by its likeness to that thing. In judgment,

---

427 Aquinas concludes: "Accordingly some hold that the meanings of these terms connote various corresponding divine effects: for they maintain that when we say God is good, we indicate God's essence together with a connoted effect, the sense being God is and causes goodness, so that the difference in these attributions arises from the difference in his effects. But this does not seem right: because seeing that an effect proceeds in likeness to its cause, we must needs understand a cause to be such before its effects are such. Wherefore God is not called wise because he is the cause of wisdom: but because he is wise, therefore does he cause wisdom" (*On the Power of God* III, q. 7, a. 6 body).

428 Aquinas explains further: "Again if when we say God is good we mean nothing more than God is and is the cause of goodness, it would follow that we could equally predicate of him the names of all the divine effects, for instance, that God is heaven since he is the cause of heaven. --Again this is clearly false if it refer to actual causality: because then we could not say that God was good, wise or the like from eternity, for he did not cause things actually from eternity" (*On the Power of God* III, q. 7, a. 6 body).

429 Aquinas continues: "If on the other hand it refer to virtual causality, so that God be called good because he is and has the power to infuse goodness; then we shall have to say that the term good signifies that power. Now that power is a supereminent likeness of its effect even as the power of any equivocal agent. Thus it would follow that the intellect in conceiving goodness is like that which is in God and is God: so that something that is in God and that is God corresponds to the notion or conception of goodness" (*On the Power of God* III, q. 7, a. 6 body).

a proper conformity between thing and the concept of that thing must exist for that judgment or proposition to be true.

Since these quiddities of the perfections observed in creatures are imperfect and improportionate likenesses of God's essence, nothing prevents God's essence from corresponding to all these various concrete concepts (concepts grounded in *modus intelligendi*), as being imperfectly represented by these perfections observed in creatures (perfections grounded in *modus essendi*). In this way, these imperfect likenesses of God are in the mind as their subject, but these likenesses are in God as the foundation of their truth since the idea that the intellect has of a thing (*res*) is not true unless that thing (*res*) conforms to the idea by its likeness to it.[430]

The intellect compares the divine essence in itself, but the intellect only sees the divine essence through many imperfect or faulty likeness of the divine essence, which are reflected by creatures as by a mirror. Although imperfect knowledge of God's essence through his effects or creatures is available, such knowledge is limited and hence there is no way to know fully what God is from his effects or through the imperfect participation of creatures in God's perfection.[431] For Aquinas each

---

430 Aquinas explains: "We must say then that all these many and diverse notions correspond to something in God of which they are likenesses. For it is plain that one form can have but one specific likeness proportionate to it: while there can be many imperfect likenesses, each one of which falls short of a perfect representation of the form. Since then, as we have proved above, the ideas we conceive of the perfections we observe in creatures are imperfect and improportionate likenesses of the divine essence, nothing prevents the same one essence from corresponding to all these ideas, as being imperfectly represented thereby. So that all these conceptions are in the mind as their subject, but in God as the foundation of their truth. For the idea that the intellect has of a thing is not true unless that thing corresponds to the idea by its likeness to it" (*On the Power of God* III, q. 7, a. 6 body). "It has already been explained that though God is absolutely one, yet these many concepts or notions are not false, because to all of them one and the same thing corresponds albeit imperfectly represented by them: but they would be false if nothing corresponded to them" (*On the Power of God* III, q. 7, a. 6, reply 4). "Since in God there is absolute unity, and multiplicity in creatures, just as God understands many creatures by one intelligible species which is his essence, while there is a manifold relationship of God to creatures: even so in our intellect which mounts up to God from the multiplicity of creatures, there must be many species having relations to one God" (*On the Power of God* III, q. 7, a. 6, reply 5).
431 Aquinas indicates that sensibles cannot lead one to know *what God is* since sensibles are effects unequal to the power of their cause, and yet through sensibles the intellect is led by sensibles to the divine knowledge that God is and similar truths: "Wherefore, if the human intellect comprehends the essence of a particular thing, for instance a stone or a triangle, no truth about that thing will surpass the capability of human reason. But this does not happen to us in relation to God, because the human

imperfect likeness reflecting the divine essence by creatures is a term, and terms predicate various notes of a subject by which the intellect understands something of that subject as by a mirror.[432] Hence Aquinas maintains that knowledge of God can be gathered from creatures until the Word, the perfect *conceptus*, returns in that day when God's very essence will be seen, and knowledge of God will no longer be gathered from creatures.[433] However, in this life perfections can be attributed to God either properly or metaphorically. Aquinas explains:

> Accordingly in creatures <u>there are certain perfections</u>
> <u>whereby they are likened to God, and which as regards the</u>

intellect is incapable by its natural power of attaining to the comprehension of His essence: since our intellect's knowledge, according to the mode of the present life, originates from the senses: so that things which are not objects of sense cannot be comprehended by the human intellect, except in so far as knowledge of them is gathered from sensibles. Now sensibles cannot lead our intellect to see in them what God is, because they are effects unequal to the power of their cause. And yet our intellect is led by sensibles to the divine knowledge so as to know about God that He is, and other such truths, which need to be ascribed to the first principle" (*SCG* I, 3).

Aquinas in discussing whether the intellect by its natural powers can see the Divine essence, Aquinas concludes: "the created intellect cannot see the essence of God, unless God by His grace unites Himself to the created intellect, as an object made intelligible to it" (*ST* I, q12, a4 body; cf. SCG IIIa, 52). Although Aquinas qualifies his position in certain instances, it is clear that in this life one cannot know God's essence in itself (*ST* I, q12, a11 body; *ST* I, q12, a11 reply 2; *ST* I, q12, a12 body; *ST* I, q12, a13 reply 1). Aquinas summarizes his position: "<u>Although by the revelation of grace in this life we cannot know of God "what He is,"</u> and <u>thus are united to Him as to one unknown; still we know Him more fully according as many and more excellent of His effects are demonstrated to us,</u> and <u>according as we attribute to Him some things known by divine revelation,</u> to which natural reason cannot reach, as, for instance, that God is Three and One" (*ST* I, q12, a13 reply 1). Although this is the case, Aquinas explains that <u>when something is predicated of God, they are predicated of his essence</u> allowing one to predicate certain perfections of God by natural reason from creatures as by a mirror (emphasis added; cf. *On the Power of God* III, q. 7, a4-a11; *ST* I, q. 13, a. 12; *ST* I, q. 13, a. 4; *SCG* I, 35; *ST* I, q. 13, a. 5; *SCG* I, 32 seqq; *ST* I, q. 13, a. 7; *ST* I, q. 28, a. 4; *ST* I, q. 32, a. 2). In regard to negative theology: "Moreover the idea of negation is always based on an affirmation: as evinced by the fact that every negative proposition is proved by an affirmative: wherefore unless the human mind knew something positively about God, it would be unable to deny anything about him. And it would know nothing if nothing that it affirmed about God were positively verified about him. Hence following Dionysius (*Div. Nom.* xiii) we must hold that these terms signify the divine essence, albeit defectively and imperfectly: the proof of which is as follows" (*On the Power of God* III, q. 7, a. 5 body).

432 Aquinas explains: "Accordingly the cause of difference or multiplicity in these expressions is on the part of the intellect, which is unable to compass the vision of that divine essence in itself, but sees it through many faulty likenesses thereof which are reflected by creatures as by a mirror. Whereof if it saw that very essence, it would not need to use many terms, nor would it need many conceptions" (*On the Power of God* III, q. 7, a. 6 body).

433 Aquinas concludes: "For this reason God's Word, which is his perfect concept, is but one: wherefore it is written (Zach. xiv, 9): In that day there shall be one Lord, and his name will be one--when God's very essence will be seen, and knowledge of God will not be gathered from creatures" (*On the Power of God* III, q. 7, a. 6 body).

thing signified do not denote any imperfection, such as being, life, understanding and so forth: and these are ascribed to God properly, in fact they are ascribed to him first and in a more eminent way than to creatures. And there are in creatures certain perfections wherein they differ from God, and which the creature owes to its being made from nothing, such as potentiality, privation, movement and the like. These are falsely ascribed to God: and whatsoever terms imply suchlike conditions cannot be ascribed to God otherwise than metaphorically, for instance lion, stone and so on, inasmuch as matter is included in their definition. They are, however, ascribed to him metaphorically by reason of a likeness in their effects (emphasis added).[434]

In contrast to Kant and Hume, knowledge of God can be gathered from creatures and something of the divine essence can be gathered from the imperfect likeness *in rebus* reflecting the divine essence as by a mirror. Further, in creatures there are perfections that do not signify any imperfection such as being, life, understanding and so forth that can be ascribed to God properly. The various terms taken to be God's effects signify one simple signification, namely the divine essence.[435] Hence it is by knowing the terms or God's effects which are the perfections found in creatures that the intellect "mounts up to God from the multiplicity

---

434 *On the Power of God* III, q. 7, a. 5 reply 8.

435 Aquinas explains: "…we have all these terms [perfections found in creatures] with one simple signification, namely the divine essence. But it must be observed that the signification of a term does not refer to the thing immediately but through the medium of the mind: because words are the tokens of the soul's impressions, and the conceptions of the mind are images of things, according to the Philosopher (Peri Herm. i)… We must say then that all these many and diverse notions correspond to something in God of which they are likenesses. For it is plain that one form can have but one specific likeness proportionate to it: while there can be many imperfect likenesses, each one of which falls short of a perfect representation of the form. Since then, as we have proved above, the ideas we conceive of the perfections we observe in creatures are imperfect and improportionate likenesses of the divine essence, nothing prevents the same one essence from corresponding to all these ideas, as being imperfectly represented thereby. So that all these conceptions are in the mind as their subject, but in God as the foundation of their truth. For the idea that the intellect has of a thing is not true unless that thing corresponds to the idea by its likeness to it. Accordingly the cause of difference or multiplicity in these expressions is on the part of the intellect, which is unable to compass the vision of that divine essence in itself, but sees it through many faulty likenesses thereof which are reflected by creatures as by a mirror" (*On the Power of God* III, q. 7, a. 6 body).

of creatures" and in this way there are many species having relations to one God."[436] Aquinas also argues that this corresponding relationship is neither univocal nor equivocal but analogical.[437] Aquinas explains the specific kind of analogical relationship that exists between God as the first efficient and formal cause and that of creatures:

> We must accordingly take a different view and hold that nothing is predicated univocally of God and the creature: but that those things which are attributed to them in common are predicated not equivocally but analogically. Now this kind of predication is twofold. The first is when one thing is predicated of two with respect to a third: thus being is predicated of quantity and quality with respect to substance. The other is when a thing is predicated of two by reason of a relationship between these two: thus being is predicated of substance and quantity. In the first kind of predication the two things must be preceded by something to which each of them bears some relation: thus substance has a respect to quantity and quality: whereas in the second kind of predication this is not necessary, but one of the two must precede the other. Wherefore since nothing precedes God, but he precedes the creature, the second kind of analogical predication is applicable to him but not the first (emphasis added).[438]

Aquinas indicates that analogy proper to the relationship between God and creature is that sort of analogy that holds between a cause and its effect by reason of the relationship that the cause and effect share. The relationship that substance has to quantity is that of being where substance precedes quantity and they are related in that the being of substance is the basis of the quantity predicated in or of substance. Likewise, God precedes creatures and the perfections that both share allow for analogical predication. For example, since God is good and has the power to infuse goodness and since creatures express an

---

436 *On the Power of God* III, q. 7, a. 6 reply 5.
437 *On the Power of God* III, q. 7, a. 7 body.
438 *On the Power of God* III, q. 7, a. 7 body.

imperfect perfection of goodness insofar as the communication of *esse* infuses goodness and since act follows being, good can be analogically predicated of both God and creatures.

In the predicated analogy, something must be true of both God and creatures and it is this relationship of imperfect participation on the part of the creature that can be predicated of God. The divine essence itself is the supereminent likeness of all things. It is by reason of this likeness that good and the like are predicated in common of God and creatures.[439] Aquinas argues that the knowledge of God taken from creatures is not simply a nominal agreement since we would only have empty expressions to which nothing would in fact correspond in reality. This in turn would make all arguments about God sophistry.[440]

Unlike Kant who held that any natural knowledge of God from experience is not possible, Aquinas would argue that an appeal to faith is required on those matters that exceed human reason, but not all divine things exceed natural reason and even with those that do so, reason is required. A Kantian nominal agreement taken to be a faith assertion would be considered an empty expression to which nothing would in fact correspond in reality since God is not an object of possible experience nor can God's existence be deduced from his effects, and this would make all arguments about God sophistry which for Aquinas would be absurd. Further, the Kantian notion makes all talk about God mere opinion and meaningless which was never Kant's intent for otherwise God as a regulative notion for moral conduct would have been non-sense. Kant's nominal agreement, making room for faith, was simply mistaken and inconsistent with Kant's intent to make the conception of God a regulative notion for moral conduct.

---

439 *On the Power of God* III, q. 7, a. 7 reply 6th obj.

440 Aquinas argues: "Since all our knowledge of God is taken from creatures, if the agreement were purely nominal, we should know nothing about God except empty expressions to which nothing corresponds in reality. Moreover, it would follow that all the proofs advanced about God by philosophers are sophisms: for instance, if one were to argue that whatsoever is in potentiality is reduced to actuality by something actual and that therefore God is actual being, since all things are brought into being by him, there will be a fallacy of equivocation; and similarly in all other arguments" (*On the Power of God* III, q. 7, a. 7 body).

In Aquinas's Commentary on the *De Trinitate of Boethius* (ca. 1257/1258-9), Aquinas asks the question whether man can arrive at a knowledge of God. Aquinas lists a number of objections why one cannot have any knowledge about God apart from revelation, but then he concludes such a position is contrary to Scripture and contrary to the various authorities. Aquinas selects Romans 1.20 without commentary. In the case of Jer. 9.24 Aquinas argues that one cannot glory in God if one cannot know God, therefore Aquinas concludes we can in fact know God if we are to glorify God. Likewise Aquinas comments on Augustine, indicating that since we have been commanded to love God, we are able to have some natural knowledge of God. Otherwise, we have been commanded to do something that is impossible. For it is impossible to love God without having some natural knowledge of God according to the opinion of Augustine. Similarly, Aquinas argues that faith presupposes natural knowledge, even as grace presupposes nature since faith is assent to what can be known.[441]

Finally, Aquinas argues that there are two ways of knowing something. First, one can sense something and thereby come to know some likeness of that thing, for example seeing a house and imagining the likeness of that house in one's imagination and being able to describe that house in some detail. Second, one can know something by apprehending something similar, for example, one can see a picture and know someone through the likeness given in that picture, or one can see an effect finding some similarity to its cause as in the case of seeing burning wood and thereby apprehending that it was fire that started the wood to catch fire. Aquinas writes:

> On the contrary, it is said in Romans 1:20: "Ever since the creation of the world his invisible nature, namely his eternal power and deity, has been clearly perceived in the things that have been made."

---

441 *ST* I, q. 2, a. 2 reply 1.

2. Jeremiah 9:24 says: "Let him who glories glory in this, that he understands and knows me." But this glory would be empty if we could not know him. Therefore we can know God.

3. Nothing is loved unless it is known, as Augustine explains. But we are commanded to love God. Therefore we can know him, for we are not ordered to do the impossible.

Reply: <u>Something can be known in two ways</u>: in one way <u>through its own form</u>, as the eye sees a stone <u>through the likeness of a stone</u>; in another way <u>through the form of something else similar to itself</u>, as <u>a cause is known through the likeness of its effect</u> and a man through the form of his image (emphasis added).[442]

Aquinas then gives two arguments. The first argument is used to demonstrate that the human intellect is limited and therefore the intellect cannot fully know or apprehend what God's essence is. However, the second argument argues that the human intellect can still know something about God through God's creation or his effects, so God is not entirely unknown to the human intellect or powers of perception. Aquinas's argument of the limitation of human reason consists of two syllogisms and a conclusion.

The first syllogism argues that our intellect is oriented to the corporeal and sensible things from which we abstract what things are, and because God's essence cannot be abstracted from sensible things we cannot know God's essence. Aquinas explains: "Because our intellect in its present state has a definite relationship to forms that are abstracted from the senses, being related to images as sight to colors, as the treatise *On the Soul* says, in its present state it cannot know God through the form that is his essence."[443] The second syllogism argues that the intellect only knows created forms and created forms are insufficient to make

---

442 *In De Trinitate of Boethius*, q. 1 a. 2 in *Faith, Reason, and Theology, Questions I-IV of the Commentary on Boethius' De Trinitate*, trans. by Armand Maurer (Toronto: Pontifical Institute of Mediaeval Studies, 1986); all citations of *In De Trinitate* from this translation unless otherwise noted.
443 Ibid.

known God's essence even if the created forms are infused or implanted, therefore the human intellect in its natural state cannot know God's infinite transcendence. Aquinas argues: "For any likeness imprinted by him in the human mind would be insufficient to make it know his essence because it infinitely transcends every created form."[444]

Therefore Aquinas explains that what God is cannot be fully understood by the human intellect because a) this knowledge of God is not accessible by the intellect through corporeal things and b) because of the natural relation of our intellect to corporeal objects or sensible things is insufficient to allow for infused knowledge of God's infinite transcendence. In this limitation of reason Kant would readily agree. Aquinas explains: "as Augustine says, God cannot be accessible to the intellect through created forms. Neither in the present life do we know God through purely intelligible forms that bear some likeness to him; this is because of the natural relation of our intellect to images mentioned above."[445] Thus Aquinas argues for the limitation of the human intellect in knowing God's transcendent nature or essence either by natural reason or even by infused divine illumination.

However, Aquinas then argues that although the human intellect is insufficient to know the transcendent essence of God, one can argue for God's existence from his effects. The demonstration argues that God's existence can be known from his effects even though God's transcendent essence cannot be fully apprehended or known. Once the basic argument is made, Aquinas then describes various characteristics of what can be known from a given effect and then applies this to God as the first efficient cause of creation. God's effects are limited in their ability to describe God's infinite transcendence, but at least they can indicate something about God's power, God's likeness, and how different God is from his creation or God's effects. Thus by this imperfect participation of the creature, by analogy one can know something about God. In knowing something about God from his effects, we thus can know that God exists by the light of natural reason.

444 Ibid.
445 Ibid.

<u>Natural Knowledge of God from God's Effects</u>

Aquinas continues to argue in his *Commentary on the De Trinitate of Boethius*, that although an effect is not entirely identical to its cause in that it falls short, one can at least know that a cause exists because the effect itself exists. There is a natural limitation in what can be known of God from existence or from his effects, and that limitation is that one can know that God exists as the First Cause of creation but not what God is in his infinite transcendence. Aquinas then explains that although an effect falls short of equality with its cause, three things can be known. The first thing that can be known is that the effect is produced from its cause. Second, that an effect acquires a likeness to its cause. Third, that an effect falls short of equality with its cause and thus a lack of perfect equality with its cause can be known. Aquinas extrapolates from certain principles found in natural philosophy and metaphysics to argue that an efficient cause produces something like itself (*omne agens agit sibi simile*) and Aquinas applies this principle to God.

God is the cause of creation and thus the effect of creation is like God who is its cause.[446] It is presupposed based upon earlier demonstration that

446 Aquinas explains in some detail: "Effects that fall short of their causes do not agree with them in name and ratio, and yet there must needs be some likeness between them, because it is of the nature of action that a like agent should produce a like action, since every thing acts according as it is in act. Wherefore the form of the effect is found in its transcendent cause somewhat, but in another way and another ratio, for which reason that cause is called equivocal. For the sun causes heat in lower bodies by acting according as it is in act; wherefore the heat generated by the sun must needs bear some likeness to the sun's active power by which heat is caused in those lower bodies and by reason of which the sun is said to be hot, albeit in a different ratio. And thus it is said to be somewhat like all those things on which it efficaciously produces its effects, and yet again it is unlike them all in so far as these effects do not possess heat and so forth in the same way as they are found in the sun. Thus also God bestows all perfections on things, and in consequence He is both like and unlike all.

Hence it is that Holy Writ sometimes recalls the likeness between Him and His creatures, as when it is said (Gen. i. 26): Let Us make man to Our image and likeness: while sometimes this likeness is denied, according to the words of Isa. xl. 18: To whom then have you likened God; or what image will you make for Him? and of the psalm: O God, who shall be like to Thee?

Dionysius is in agreement with this argument, for he says (Div. Nom. ix.): The same things are like and unlike to God; like, according as they imitate Him, as far as they can, Who is not perfectly imitatable; unlike, according as effects fall short of their causes.

However, according to this likeness, it is more fitting to say that the creature is like God than vice versa. For one thing is like another when it possesses a quality or form thereof. Since then what is in God perfectly is found in other things by way of an imperfect participation, that in which likeness is observed is God's simply but not the creature's. And thus the creature has what is God's, and therefore is

since creation exists, God who is its cause must likewise exist because an effect follows from its cause and cannot exist without a cause. Therefore from God's effects, that fall short of equality with God, three things can be known in regard to God's existence. First, one can know more perfectly God's power in producing the effect, and since the power is a supereminent likeness of God's effect, one can conceive the likeness of God through his effects. Second, by knowing the resemblance between God's effects and God as cause, one knows something about God's invisible nature and greatness. Third, by knowing how God's effects fall short of God as cause, one gains an ever-growing knowledge of God's transcendent deity through separation and negation of God's effects from God as cause.[447]

Aquinas concludes the question related to Romans 1.20 with the following: "The human mind receives its greatest help in this advance of knowledge when its natural light is strengthened by a new illumination, like the light of faith and the gifts of wisdom and understanding, through which the mind is said to be raised above itself in contemplation, inasmuch as it knows that God is above everything it naturally comprehends."[448] Aquinas concludes that the human mind, although it can know something about God from his creation, the natural light of the human intellect can be strengthened by the

---

rightly said to be like God. But it cannot be said in this way that God has what belongs to His creature: wherefore neither is it fitting to say that God is like His creature; as neither do we say that a man is like his portrait, although we declare that his portrait is like him.

And much less properly can it be said that God is assimilated to the creature. For assimilation denotes movement towards similarity, and consequently applies to one that receives its similarity from another. But the creature receives from God its similarity to Him, and not vice versa. Therefore God is not assimilated to His creature, but rather vice versa." *SCG* I, 29.

447 Aquinas argues: "Now we can consider from three points of view the relation in an effect that falls short of equality with its cause: with respect to the coming forth of the effect from the cause, with respect to the effect acquiring a likeness to its cause, and with respect to its falling short of perfectly acquiring it. So the human mind advances in three ways in knowing God, though it does not reach a knowledge of what he is (*quid est*), but only that he is (*an est*). First, by knowing more perfectly his power in producing things. Second, by knowing him as the cause of more lofty effects which, because they bear some resemblance to him, give more praise to his greatness. Third, by an ever-growing knowledge of him as distant from everything that appears in his effects. Thus Dionysius says that we know God as the cause of all things, by transcendence and by negation" (*In De Trinitate of Boethius* q. 1, a. 2; emphasis added).

448 *In De Trinitate of Boethius* q. 1, a. 2.

illumination that comes through the light of faith and the gifts of wisdom and understanding.

In contrast, for Kant *a priori* knowledge extends only as far as a *range of experience* via *empirical intuition of appearance* in virtue of which a knowledge of God would not be considered cognitive, but this denial of a knowledge of God and particularly of God's existence also has the positive advantage according to Kant of making room for faith: "I have therefore found it necessary to deny knowledge [of God], in order to make room for faith. The dogmatism of metaphysics, that is, the preconception that it is possible to make headway in metaphysics without a previous criticism of pure reason, is the source of all that unbelief, … which wars against morality."[449]

The negative effect of Kant's positive advantage in making room for faith denied the possibility of speculative knowledge and demonstration of God's divine action and divine causality thus limiting speculative knowledge of God's existence since *a priori* knowledge is confined to the range of experience *via* the empirical intuition of appearances. Divine action, divine causality, and God's existence were outside the range of the empirical intuition of appearances for Kant, and therefore beyond the limits of pure reason. It is this fundamental shift in epistemology and the Kantian understanding of causality, universality, and necessity that consequently led to an inability to justify a natural knowledge of God's existence from efficient causality and necessity since such knowledge is apprehended inductively from God's effects. However, such effects even in the Kantian model are empirical intuitions of possible experience, but pure reason for Kant limits one's ability to deduce from an effect existing as an empirical intuition of possible experience a cause not available as an empirical intuition.

Although Aquinas agrees with the limitations of human reason to obtain a natural knowledge of God and often repeats that "God can be thought not to be not because of God's imperfections or uncertainty

---

449 *Critique of Pure Reason*, translated by N. Kemp Smith (Macmillan, London: 1929), 29.

but because of the weakness of our intellect," and yet Aquinas maintains that because of this limitation, "one cannot behold God Himself *except through His effects that lead to a knowledge of His existence through reasoning*" (SCG I 11.4; emphasis added).[450] While Kant excludes the possibility of a natural knowledge of God from his effects, Aquinas argues that such knowledge is only possible through his effect and by the light of faith the human intellect can be strengthened or perfected or raised above itself in contemplation with a new illumination through the light of faith and wisdom. Through this divine illumination, the mind is said to be raised above itself and arrives at a knowledge that God is above everything that the intellect can naturally comprehend on its own.

The Prima Via: The Argument from Motion

In the first way Aquinas argues for God's existence from motion and Gilson noted that the first mover given in the *Compendium Theologiae* indicates that Aquinas was interpreting the *efficacy* of the Prime Mover as that of an efficient cause.[451] Aquinas argues in the first way that something cannot be moved except insofar as it is in potentiality to the actuality toward which it is moved, and something can only effect that motion, insofar as it is in actuality. Aquinas then indicates that this is the case since to effect motion is to lead something from potentiality into actuality, but this movement cannot occur except by some agent or being or efficient cause that is in actuality, and being in actuality the efficient cause or agent of motion thereby moves and alters some potency toward some actuality.[452]

---

450 Aquinas explains: "Creatures fail to represent their creator adequately. Consequently, through them we cannot arrive at a perfect knowledge of God. Another reason for our imperfect knowledge is the weakness of our intellect, which cannot assimilate all the evidence of God that is to be found in creatures. It is for this reason that we are forbidden to scrutinize God's attributes overzealously in the sense of aiming at the completion of such an inquiry, an aim which is implied in the very notion of overzealous scrutiny. If we were to act thus, we would not believe anything about God unless our intellect could grasp it. We are not, however, kept from humbly investigating God's attributes, remembering that we are too weak to arrive at a perfect comprehension of Him" (*De Veritate* I, q. 5, a. 2, reply 11).

451 Dennis Bonnette, *Aquinas' Proofs for God's Existence, St. Thomas Aquinas On: "The Per Accidens Necessarily Implies the Per Se"*, (Netherlands: Martinus Nijhoff, 1972), 105.

452 Aquinas argues from motion that "nothing is moved except insofar as it is in potentiality with

Aquinas thus argues that it is impossible for something to be simultaneously in potentiality and in actuality with respect to the same thing, and likewise drawing upon modal necessity, Aquinas maintains that it is impossible for something to be both mover and the thing moved in the same way and with respect to the same thing thus concluding from motion that movement from potency to act, everything that is moved is moved by another.[453] It is here in the argument from motion that Aquinas appeals to natural necessity and reduces the arguments to the first principle of speculative reason, i.e. the law of contradiction. Something cannot both be and not be at the same time and in the same respect in potency and in act. In the second way Aquinas will argue that no effect can be a cause of itself otherwise it would have to be prior to itself and from this Aquinas argues against infinite regress in efficient causality.

Since Aquinas's notion of motion is drawn from Aristotle's *Physics*, it is necessary to review Aristotle's conception of motion. Aristotle explains why there cannot be an infinite regress in motion and why it is necessarily the case that a first mover not be moved by another but moved by itself. By analogy Aristotle describes the motion of animate creatures, an animate being has in itself its own cause of motion since the animate being is not moved by another, and being in motion, an animate being is able to move another.[454] For example, man being in

---

respect to that actuality toward which it is moved, whereas something effects motion, insofar as it is in actuality…After all, to effect motion is just to lead something from potentiality into actuality. But a thing cannot be led from potentiality into actuality except through some being that is in actuality in a relevant respect; for example, something that is hot in actuality—say, a fire—makes a piece of wood, which is hot in potentiality, to be hot in actuality, and it thereby moves and alters the piece of wood" (*ST* I, q. 2, a. 3).

453 Aquinas indicates: "It is impossible for something to be simultaneously in potentiality and in actuality with respect to the same thing; rather, it can be in potentiality and in actuality only with respect to different things. For what is hot in actuality cannot simultaneously be hot in potentiality; rather, it is cold in potentiality. Therefore, it is impossible that something should be both mover and moved in the same way and with respect to the same thing, or, in other words, that something should move itself. Therefore, everything that is moved must be moved by another" (*ST* I, q. 2, a. 3).

454 Aristotle argues: "The fact is evident above all in the case of animate beings; for it sometimes happens that there is no motion in us and we are quite still, and that nevertheless we are then at some moment set in motion, that is to say it sometimes happens that we produce a beginning of motion in ourselves from within ourselves, without anything having set us in motion from without. We see nothing like this in the case of inanimate things, which are always set in motion by something else from without: the animal, on the other hand, we say, moves itself; therefore, if an animal is ever in a state of absolute

motion can move a stick by hand. The stick is thus moved by another and the stick will not move anything unless it is moved by another being itself an inanimate object. The last motion, which is that of the stick, will not move anything without the first motion which is the motion of the animate being.

Likewise, Aristotle affirms that generation and corruption in the universe or motion from potency to act requires a series of *movers and moved* in the universe and thus that which is moved must be moved by something, and something is either moved by something external to itself or not moved except by itself. In the former case, there must be some first mover that is not itself moved by anything else since in an infinite series there is no first term, and since the first mover is of the kind that it has no need to be moved by another since it is its own self-movement, there is no need for another mover. If it is the case that everything that is in motion is moved by something, and the first mover is moved but not by anything else, it must be moved by itself.[455] Aristotle argues:

> Every mover moves something and moves it with something, either with itself or with something else: e.g. a man moves a thing either himself or with a stick, and a thing

---

rest, we have a motionless thing in which motion can be produced from the thing itself, and not from without. Now if this can occur in an animal, why should not the same be true also of the universe as a whole? If it can occur in a small world it could also occur in a great one; and if it can occur in the world, it could also occur in the infinite; that is, if the infinite could as a whole possibly be in motion or at rest" (Ari: *Physics* VIII, 252b17-252b28).

455 Aristotle indicates: "Now this may come about in either of two ways, either not because of the mover itself, but because of something else which moves the mover, or because of the mover itself. Further, in the latter case, either the mover immediately precedes the last thing in the series, or there may be one or more intermediate links: e.g. the stick moves the stone and is moved by the hand, which again is moved by the man; in the man, however, we have reached a mover that is not so in virtue of being moved by something else. Now we say that the thing is moved both by the last and by the first of the movers, but more strictly by the first, since the first moves the last, whereas the last does not move the first, and the first will move the thing without the last, but the last will not move it without the first: e.g. the stick will not move anything unless it is itself moved by the man. If then everything that is in motion must be moved by something, and by something either moved by something else or not, and in the former case there must be some first mover that is not itself moved by anything else, while in the case of the first mover being of this kind there is no need of another (for it is impossible that there should be an infinite series of movers, each of which is itself moved by something else, since in an infinite series there is no first term)—if then everything that is in motion is moved by something, and the first mover is moved but not by anything else, it must be moved by itself" (emphasis added; Ari: *Physics* VIII, 256a4-256a21).

is knocked down either by the wind itself or by a stone propelled by the wind. <u>But it is impossible for that with which a thing is moved to move it without being moved by that which imparts motion by its own agency; but if a thing imparts motion by its own agency, it is not necessary that there should be anything else with which it imparts motion, whereas if there is a different thing with which it imparts motion, there must be something that imparts motion not with something else but with itself, or else there will be an infinite series. If, then, anything is a mover while being itself moved, the series must stop somewhere and not be infinite.</u> Thus, if the stick moves something in virtue of being moved by the hand, the hand moves the stick; and if something else moves with the hand the hand also is moved by something different from itself. So when motion by means of an instrument is at each stage caused by something different from the instrument, this must always be preceded by something else which imparts motion with itself. <u>Therefore, if this is moving and there is nothing else that moves it, it must move itself. So this reasoning also shows that, when a thing is moved, if it is not moved immediately by something that moves itself, the series brings us at some time or other to a mover of this kind</u> (emphasis added).[456]

Aristotle argues that the first mover has motion in itself and is moved by no other and Aristotle indicates that motion is continuous and in contact, and as such, motion that is primary and continuous is the motion that is imparted by the first mover.[457] The universe itself

456 Ari: *Physics* VIII, 256a22-256b2.

457 Aristotle explains: "For there must be three things--the moved, the mover, and the instrument of motion. Now the moved must be in motion, but it need not move anything else; the instrument of motion must both move something else and be itself in motion (for it changes together with the moved, with which it is in contact and continuous, as is clear in the case of things that move other things locally, in which case the two things must up to a certain point be in contact); and the mover--that is to say, that which causes motion in such a manner that it is not merely the instrument of motion--must be unmoved. Now we see the last things, which have the capacity of being in motion, but do not contain a motive principle, and also things which are in motion but are moved by themselves and not by anything else: it is reasonable, therefore, not to say necessary, to suppose the existence of the third term also, that which causes motion but is itself unmoved" (Ari: *Physics* VIII, 256b28-257a31). Again Aristotle

is moved by the first mover since the universe is placed in motion by another since a first mover has been demonstrated to be required to move any continuous series of motion.

Aristotle thus argues that if something is moving and there is nothing else that moves it, it must move itself. Since this is the case, when a thing is moved, if it is not moved immediately by something that moves itself, the series brings us to a first mover that moves itself for it is impossible that there should be an infinite series of movers, each of which is itself moved by something else, since in an infinite series there is no first term. For without a first there can be no intermediate, if no intermediate then no final movement and therefore an infinite series of motion is impossible because there must necessarily be a first motion for there to be any motion within a series. Therefore there can be no infinite series. Necessarily there can only be a series of movers and moved since it is necessary that there is a first mover that moves itself and is the source of motion in others.

Thus Aristotle continues and argues for a simple eternal unmoved mover that is the first cause of motion. A first mover necessarily exists since motion itself is eternal and continues and thus requires an eternal mover that is not moved by another but that is moved only by itself since an infinite regress of movers is impossible while motion itself is continuous.[458] Since an infinite series of movers is impossible and motion is continuous, motion itself must terminate in a first mover that is itself unmoved by another but is its own cause of motion and the cause of motion in others.

Likewise, Aquinas argues in the *Prima Via* that there cannot be an infinite regress of motion from potency to act by an infinite series of movers or agents since there would be no first mover and if this were

---

indicates: "We must consider whether it is or is not possible that there should be a continuous motion, and, if it is possible, which this motion is, and which is the primary motion; for it is plain that if there must always be motion, and a particular motion is primary and continuous, then it is this motion that is imparted by the first mover, and so it is necessarily one and the same and continuous and primary" (Ari: *Physics*, VIII 260a27-260b14).

458 Ari: *Physics* VIII, 257a32-260a10.

the case nothing would effect motion since there would be no first cause of motion. Of course, what is assumed in this infinite regress argument by Aquinas is Aristotle's notion of continuous motion and the arguments expressed by Aristotle for a first unmoved mover that can only be moved in itself and is the cause of motion in others.

Since motion is an imperfect act moving from potency to act and is a certain beginning of the perfect act in that which is moved, motion is fulfillment of what is potential as potential and thus motion is a movement from potency to act.[459] Since motion is therefore a mean between two extremes, potency and act, the intelligibility [*ratio*] of cause and effect is implied in motion. This is the case because a thing is only reduced from potency to act by some agent cause. From this motion belongs the predicaments of action and passion, since these two predicaments are taken from the intelligibility of an agent cause and effect. The predicament of action is used to distinguish an agent from its effect, and the predicament of passion is used to distinguish between the patient and its cause. The predicaments of action and passion are not entirely external to the subject since action is in the agent as its principle and passion is in the patient as its term.[460] Thus

---

459 Aristotle defines motion generally as follows: "the fulfillment of what is potentially, as such, is motion--e.g. the fulfillment of what is alterable, as alterable, is alteration; of what is increasable and its opposite, decreasable (there is no common name for both), increase and decrease; of what can come to be and pass away, coming to be and passing away; of what can be carried along, locomotion" (Ari: *Physics* III, 201a10-201a14).

460 *In Meta.*, V, 9.890-92; *In Phys.*, III, 4. Kossel describes the relation between motion and the *predicaments* of action and passion as follows: "St. Thomas says that action and passion are objectively identified with the motion in the patient. Motion considered as from the agent is action; motion considered as in the patient is passion. However, action and passion do not signify motion itself. Motion is simply an imperfect act which reason views as a medium between two terms in a patient, one term being potency and the other act with regard to any particular motion. To bring the potential to act, an efficient cause is required; and it is precisely these notions of cause and effect which action and passion signify (*In Phys.* III, 5; *In Phys.*, V, 2 and 3; *In Meta.* XI, 9.2312-13; *SCG* II, 57; *In de Anima* I, 6.726; *ST* I, q. 28, a. 3 ad1; *ST* I, q. 41, a. 1 ad 2; *ST* I-II, q. 110, a. 2). Hence St. Thomas says that even where there is causality without motion, there is action and passion, for there is influence of cause on effect... Even in the purest type of transitive action, creation, which presupposes neither matter nor motion but only the actuality of the agent, there is action and passion (*In Phys.*, VIII, 2; *In Meta.* VII, 7.1417ff; *De Veritate* III, q. 26, a. 1; *De Pot.* 3.2, 3-4; *Quodlibet* I, a. 6; *Quodlibet* VII, a. 9; *ST* I, q. 45, a. 1-4; *ST* I-II, q. 113, a. 7). What happens, then, when an agent brings about an effect, with or without [local motion]? In its widest sense, to act is to communicate; the agent communicates something of its actuality to the patient (De. Pot., 2.1). This is not to be understand as a transfer. Neither a substantial nor an accidental form of the agent passes physically to the patient; the agent does not lose something

Aquinas can argue from natural modal necessity of act and potency essentially in the subject and patient to a first mover as well as to a first efficient cause and a necessary being. Hence, Aquinas argues from action in an efficient causal agent to passion in an effect. The reason for associating the first efficient cause with the Prime Mover is that act precedes potentiality since nothing is reduced to act except by a being in act who is the cause of motion in others.

Aquinas intentionally separates the arguments making a distinction between the argument from motion in the *prima via*, the argument from efficient causality in the *secunda via*, and the argument from possibility and necessity in the *tertia via*, but these arguments follow certain common principles. The first principle is action follows being, that is, action follows upon the mode of actuality in an agent. Thus an effect, like motion from potency to act, follows from the *esse* (the act of existing) or being of its cause or mover. Aquinas explains: "Nothing acts except according as it is in act: wherefore action follows upon the mode of actuality in the agent; and consequently it is impossible for the effect that results from an action to have a more excellent actuality than that of the agent, although it is possible for the actuality of the effect to be more imperfect than that of the active cause, since action may be weakened on the part of that in which it terminates."[461]

This first principle is the basis for the following two principles because a causal agent in acting according to its own being generates effects similar to itself and is able from its own being to communicate

---

when it acts (*De Pot.* 2.1; SCG III, q. 69, a. 9; *ST* I, q. 115, a. 1). Rather the agent produces its like, either by reducing the potential in a pre-existing patient to act or by creation. The likeness will be there because the efficacy of a cause reaches as far as its actuality; it can cause what it is. What is man can cause man; what is hot can cause heat. God is being and can cause being; and since he causes being, his activity presupposes no passive potentiality in some subject, but only his active potency which is his being. (*Quastio de Anima* a. 12; *De Pot.* 1.3; *De Pot.* 3.7 ad 10; *De Pot.* 3.8; *ST* I, q. 25, a. 2; *ST* I, q. 42, a. 6; *ST* I, q. 44, a. 1 and 2; *ST* I, a. 45, a. 2, 5 and 8; *ST* I, a. 65, a. 4; ST I-II, a. 63, q. 1).... Only when the operation of a thing is its esse, is the essence or substantial form the immediate principle of operation; and this is true only of God (In Sent., d. 3, 4.2; Quest. De An. a.12; Quodlibet X, a.5)." See Clifford G. Kossel, S. J. "St. Thomas's Theory of the Causes of Relation", *The Modern Schoolman* vol. 25, no. 3 (March 1948), 155, 162-167). Also see Bernard Lonergan, "St. Thomas's Theory of Operation," *Theological Studies*, III (September, 1942), 375-83.
461 *SCG* I, 28.

existence to a given effect. For example, fire in being fire can both produce something like itself, i.e., heat, and can communicate existence to its effect such as wood that burns where the burning fire is the existence communicated by the efficient cause of fire. Lightening striking wood produces fire because lightening itself has a quality in itself that generates fire.

Thus the second principle used by Aquinas is that every efficient cause produces something like itself (*omne agens agit sibi simile*).[462] For example, the idea of a picture in the mind of an artist is produced on canvass, fire generates fire, man generates man, and a grain of wheat produces wheat.[463] An effect is therefore like its cause because an effect follows from the *esse* of its cause. Aquinas maintained that although in the case of effects that fall short of their cause they do not fully agree or express their cause, yet there is some likeness between the cause and its effect.[464] Aquinas explains: "It is of the nature of action that an agent produces something like itself, since each and everything acts insofar as it is in act."[465] *Esse* is the act of existing, which brings us to the third principle known inductively from nature.

The third principle is the communication of *esse* from a given cause to its effect through the process of becoming, which is a process of movement or actualization from something that is potentially the case to something that becomes actually the case. In this movement from potentiality to actuality, there is a communication of being (*esse*) from one thing to another. For example, an apple seed is potentially an apple tree where like generates like and in the process of becoming or *entelechy*, the apple seed becomes an apple tree and thereby the act of existing of the apple tree is potentially in the seed.

In the process of becoming, the *esse* of the seed communicates *esse* to the tree for every cause generates the *esse* or *existence of that effect* to

---

462 Hans Meyer, *The Philosophy of St. Thomas Aquinas*, trans. by Rev. Frederic Eckhoff, (London: B. Herder Book Co., 1948), 264.
463 Ibid., 264.
464 *SCG* I, 28.
465 *SCG* I, 28.

that effect, for without a cause communicating existence to an effect there could be no effect for being is not produced by non-being. In other words, a seed does not communicate its own act of existence to the tree otherwise the tree would be a seed, but the seed communicates the tree's own act of existence insofar as the tree becomes a tree otherwise the tree would not be.

According to Wippel, Aquinas explains that the "the form of an effect must be found in some way in its cause."[466] Meyers explains: "In the process of becoming, the being hands on its own structure and its peculiarity and its mode of activity, it reproduces its properties in other beings, not as if the cause entered into the essence of the effect but in such a way that a peculiar effect proceeds from a peculiar cause... the process of becoming is a process of actualization, the process that Aristotle called entelechy, which is nothing else than the form intent on producing a definite type. St. Thomas incorporated this thought in his natural philosophy [and his metaphysics] and utilized it in his theory of form originating from the potency of matter."[467]

Meyers indicates that this third principle was used by Averroës, St. Albert, and St. Bonaventure as developed by Aristotle.[468] For Aristotle and Aquinas, the principle of *entelechy* is expressed in and through nature, is inductively determined by observing nature, and therefore can be used as a self-evident principle, once understood, in arguments for the existence of God. This third principle of actualization was a movement from potentiality to what is actual, that is, from what potentially exists to what will in fact exist through the process of actualization or becoming.

Since motion is an imperfect act moving from potency to act and is a certain beginning of the perfect act in that which is moved,

466 John F. Wippel, "Thomas Aquinas on Our Knowledge of God and the Axiom that Every Agent Produces Something Like Itself", *Philosophical Theology: Reason and Theological Doctrine*, vol. 74 in the *Proceedings of the American Catholic Philosophical Association*, ed. Michael Baur (NY: Fordham University, 2000), 84.

467 Hans Meyer, *The Philosophy of St. Thomas Aquinas*, trans. by Rev. Frederic Eckhoff, (London: B. Herder Book Co., 1948), 268.

468 Ibid., 268.

motion is fulfillment of what is potential as potential and thus motion is movement from potency to act. Since motion is a mean between two extremes of potency and act, the intelligibility [*ratio*] of cause and effect is implied in motion.[469] This is the case because a thing is only reduced from potency to act by some agent cause. For this reason the first mover can be understood to be the first efficient cause of coming-to-be and Aquinas can interpret the efficacy of the Prime Mover as that of an efficient cause.[470]

Therefore Gilson can appropriately suggest that "St. Thomas' argument for a first mover given in the *Compendium Theologiae* suggests that he 'was interpreting the efficacy of the Prime Mover as that of an efficient cause.'"[471] Bonnette explains that this is acceptable insofar as the first mover is understood to be *the efficient cause of coming-to-be.*[472] Thus Aquinas can argue from motion to a first mover and from a given cause and effect found in movement from potency to act to the first efficient cause. Specifically the *Prima Via* argument given by Aquinas is the following:

> <u>Nothing</u> is moved <u>except</u> insofar as it is in potentiality
> with respect to that actuality toward which it is moved,

---

469 *In Physics* III, 5, 324.

470 Aquinas explains this relationship in stating: "That which is the first of beings, must needs be the cause of the things that are: for if they were not caused they would not be set in order thereby, as have already proved. Not between act and potentiality there is this order, that, although in the one and the same thing which is sometimes in potentiality and sometimes in act, potentiality precedes act in point of time, whereas act precedes by nature; nevertheless, speaking simply, act must needs precede potentiality, which is evidenced by the fact that potentiality is not reduced to act save by a being in act. But matter is a being in potentiality. Therefore God Who is pure act must needs be simply prior to matter, and consequently the cause thereof. Therefore matter is not necessarily presupposed for His action... Now God is the cause of all things as we have proved [in *SCG* 15]... Divine Scripture confirms this truth, saying (Gen. 1.1): In the beginning God created heaven and earth. For to create is nothing else than to bring something into being without preadjacent matter... Hereby is refuted the error of the ancient philosophers who asserted that matter has no cause whatever, because they observed that in the action of particular agents something is always preadjacent to action: whence they drew the opinion common to all that from nothing naught is made. This is true in particular agents. But they had not yet arrived at the knowledge of the universal agent, which is the active cause of all being, and of necessity presupposes nothing for His action" (*SCG* II, 16). Aquinas draws a relationship between motion from potency to act and from non-being to being by an agent cause taken to be a first mover, first efficient cause, or a necessary being drawing from the essential principles of being and becoming.

471 Dennis Bonnette, *Aquinas' Proofs for God's Existence, St. Thomas Aquinas On: "The Per Accidens Necessarily Implies the Per Se"*, (Netherlands: Martinus Nijhoff, 1972), 105.

472 Ibid., 105.

whereas <u>something effects motion insofar</u> as it is in actuality in a relevant respect. After all, to effect motion is just to lead something from potentiality into actuality. But a thing <u>cannot</u> [modal necessity] be led from potentiality into actuality except through some being that is in actuality in a relevant respect;... But it is <u>impossible</u> for something to be simultaneously in potentiality and in actuality with respect to same thing; rather, it can be in potentiality and in actuality <u>only with</u> respect to different things... Therefore, it is <u>impossible</u> that something should be both mover and moved in the same way and with respect to the same thing, or, in other words, that something should move itself. Therefore, everything that is moved <u>must be</u> moved by another.[473]

Aquinas using modal necessity argues that a) it is necessarily the case that nothing can be moved except insofar as it in potentiality to that actuality toward which it is moved, b) it is necessarily the case that something effects motion insofar as it is in actuality itself,[474] c) it is necessarily the case that a thing cannot be led from potentiality to actuality except by some being already in actuality, d) it is necessarily the case, or it is impossible for something 'x' to be simultaneously in potentiality and in actuality with respect to the same thing 'y' based on the law of contradiction. Hence the opposite is necessarily the case, something 'x' can be in potentiality and in actuality only with respect to numerically different things (y and z).

This is a necessary modal conditional, if 'x' then necessarily 'y'. If 'x' is in *potentiality* to 'y' then 'x' necessarily cannot be in *actuality* to 'y' without violating the law of contradiction. If 'x' is in *potentiality* to 'y' then 'x' can be in *actuality* to 'z'. Therefore Aquinas concludes that it is necessarily the case that something 'x' cannot be both a mover and moved in the same way and with respect to the same thing 'y' but 'x'

---

473 *ST* I, q. 2, a. 3.
474 This reflects modal absolute natural necessity since Aristotle's physics accepts these as necessary and universal principles in the nature of contingent things in this world.

can be a mover and moved in respect to different things otherwise the law of contradiction would be violated.

Note the argument is intentionally reduced to the law of contradiction. Each premise is so reduced and by reducing the premise and conclusions to the law of contradiction, the argument and the conclusion itself is necessary. Hence Aquinas concludes that if 'x' is the mover and the moved, 'x' cannot both be in potentiality as mover and in actuality as moved with respect to itself at the same time and in the same respect. The law of contradiction is not simply deductive drawing on logical necessity, but as a modal proposition it inductively draws from the order of nature where something cannot both be in potentiality as a mover and in actuality as moved with respect to itself. It is impossible (necessarily so since the consequent necessarily follows from the antecedent condition) for a thing moving from potency to act to move itself since it is impossible for a being to be at the same time and in the same respect in potentiality and in actuality, for it is either potential or actual and necessarily not both simultaneously.

A being in potency is in potency to being and not being, and if being is in act then it is no longer in potency to being. Since the one and the same thing cannot be numerically non-being and being, something cannot move itself from non-being to being. An animate creature moves itself by a part being in potency and a part in act, but Aquinas is speaking specifically of motion in terms of movement from potency to act of numerically the same thing. Aquinas begins the argument by stressing "It is certain, and obvious to the senses, that in this world some things are moved. But everything that is moved is moved by another." The self-evident and certain observation attained by the senses is thus demonstrated by Aquinas's modal argument, and Aquinas is therefore able to use his conclusion in a modal conditional for the next argument:

> If, then, that by which something is moved is itself moved,
> then it, too, must be moved by another, and that other by still

another. But this <u>does not</u> go on to infinity. For if it did, then there <u>would not</u> be any first mover and, as a result, <u>none</u> of the others would effect motion, either. For secondary movers effect motion <u>only</u> because they are being moved by a first mover, just as a stick <u>does not</u> effect motion <u>except</u> because it is being moved by a hand. <u>Therefore, one has to arrive at</u> some first mover that is <u>not being</u> moved by anything. And this is what everyone takes to be God.[475]

If that by which something is moved is itself moved, then it is necessarily the case that an intermediary is moved by another, and that intermediary moved by still another because it is necessarily the case that everything moved from potency to act or from non-being to being is moved by another as demonstrated by the preceding argument. If 'c' moved 'd' and if 'c' is itself moved by 'b', it is necessary for 'c' to be moved by 'b' and 'b' to be moved by another. However, 'd' being moved by 'c' and 'c' by 'b' cannot go on to infinity in an infinite series since nothing from nothing comes. Nothing moves from potency to act except by a being in act hence in an ordered series of movers and moved no secondary mover can move itself and therefore must be moved by another.

Since nothing can move itself and must be moved by another within this series from potency to act or non-being to being as demonstrated, if there was no first mover none of the others in the series could effect motion because <u>secondary movers can only (necessarily) effect motion if they are moved by a first mover that has in itself the initial cause of motion</u>. Given that all secondary movers are touched by its predecessor mover or are connected to one another in the given series by a continuous and sequential act of motion, the motion effected by the first mover effects motion in each of the secondary movers. Hence 'd' is moved by 'c' and 'c' is moved by 'b' and 'b' is necessarily moved by the first mover 'a' because it is necessarily the case that 'b' and 'c' as

---

475 *ST* I, q. 2, a. 3.

secondary or intermediate movers are moved by a first mover 'a' that is not moved by another but moved in itself and is able to move another. Iff 'a' moves 'b' and 'b' moves 'c', 'c' and 'd' are necessarily moved by 'a' because the consequent is implied in and necessarily follows from the antecedent condition that in an ordered series there must necessarily be a first mover that moves of itself and is the cause of motion in the others within that given series. [476]

Let p be "a moves b, b moves c, c moves d (in a single ordered series at time t in series s)." Let q be "b, c, and d are moved by a (at that same t,s)." Aquinas' principle of per se causal order makes $p \rightarrow q$ analytically true for that series: if the chain originates with a and is linear through $b \rightarrow c \rightarrow d$, then a is the mover of each of b, c, d in that order.

To reach a necessity claim ($\Box q$) from the mere possibility of p ($\Diamond p$), we appeal to a de re essentiality: being moved-by-a is taken as an essential dependence of the secondary movers in that series—i.e., whenever any of b, c, d are moved (in any accessible world/time for that series), they are moved-by-a. Think of the first x at singularity, for it to have come into existence it had to have been acted upon by some y for x to have come into existence. But for that y to have existed, some z must necessarily have existed. It is a causal chain, z caused y, and y caused x from which the universe emerged.

---

476 Gilson explains, "When a series of movers and things moved are ordered, that is, when they form a series where each one moves the next, it is inevitable that, if the first mover disappeared or ceased to move, none of the rest would any longer be either a mover or moved [necessarily]. It is the first mover, indeed, which confers the power of moving on all the others. Now if we have an infinite series of movers and things moved, there will no longer be [necessarily] a first mover and all are intermediate movers. Therefore, if the action of the first mover is wanting, [necessarily] nothing will be moved and there will be [necessarily] no movement in the world" (Necessity added where appropriate; Etienne, Gilson, *The Christian Philosophy of St. Thomas Aquinas*. Translated by L. K. Shook. (New York: Random House, 1956), 62).

1. The Formal Modal Argument

Definitions (at a fixed time t and series s):

p := MovesChain(a,b,c,d,t,s)

q := MovedBy(a,b,t,s) ∧ MovedBy(a,c,t,s) ∧ MovedBy(a,d,t,s)

Premises:

- (A1) □(p → q)
- (Ess) □ ∧_{x∈{b,c,d}} ∀t*∀s* [Moved(x,t*,s*) → MovedBy(a,x,t*,s*)] (de re essentiality)
- (Poss) ◇p

Derivations:

- From (A1) and (Poss): ◇q
- From (Ess): for each x∈{b,c,d}, □[Moved(x) → MovedBy(a,x)] (conditional necessity)

Conclusion (de re):

- □_de re q i.e., for x∈{b,c,d}, □[Moved(x) → MovedBy(a,x)]

2. De Re Necessity (Semantics)

Necessity de re for a dependence R on an entity x at (t,s) can be expressed as: R holds of x at (t,s) and, for any (t*,s*) where x exists and the dependence is instantiated (i.e., x is moved as a secondary mover), R holds there as well. This matches the "necessary for entity q ... at t in s iff (a) q at (t,s) and (b) for any t*,s* where x exists, R holds". Formally: □∀t*∀s*[(E(x,t*,s*) ∧ Moved(x,t*,s*)) → MovedBy(a,x,t*,s*)].

3. Metaphysical Grounding (Act/Potency)

A thing in potency cannot actualize itself to act in the same respect at the same time. Hence, whatever is moved (as secondary) is moved by something already in act. Thus, if b, c, d are secondary movers in the series rooted at a, whenever they are moved, they are moved-by-a. This yields the de re essential dependence required to uplift ◇p to a de re □ for q.

Given that 'a moves b, b moves c, c moves d' is (p) and 'b, c, d are moved by a' is (q) where p implies q, then 'possibly p' implies 'necessarily q'. The *ad absurdum* argument amounts to the following rule of inference:

$$\frac{p \supset q}{Mp \supset Nq}$$

Another formulation: Where 'M' is a shorthand for 'possibly' and 'N' is shorthand for 'necessarily'. The material implication $\supset$ indicates 'if ... then ...'. The same necessary condition can be expressed as follows using contemporary *de re* modal semantics:

It is necessary for entity q to be (or impossible for q not to be) at moment t in series s if and only if

a. q exists at t in s, and

b. for any moment t^* and any possible series s^* such that q exists at t^* in s^*, q occurs at t^* in s^*

Again the reason why the antecedent implies and necessarily follows from the consequent or why it is the case that 'if secondary movers are moved, the secondary movers are necessarily moved by a first mover' is because it is impossible (necessarily) for a thing moving from potency to act to move itself since it is impossible for a being to be at the same time and in the same respect in potentiality and in actuality to being, for it is either potential or actual and necessarily not both simultaneously. Hence something in potency cannot be moved to act except by a being in act, and therefore something that does not have in itself its own cause of motion cannot cause motion in itself nor cause motion in another. The implication being, the cause of existence cannot be existence itself, it must necessarily be that by which existence was moved into existence or caused existence. Matter cannot cause itself, it must necessarily be caused. This means that abiogenesis is virtually impossible. Likewise, some x causing itself from which the universe came into existence is necessarily impossible since it must be acted upon by another.

Likewise, being cannot be caused by non-being or motion from non-motion *nihil ex nihilo fit* and something cannot be both non-being and being at the same time and in the same respect therefore something that does not have in itself its own cause of motion cannot cause motion in itself or in another. It follows from this that there must necessarily be something that has motion in itself and that is the cause of motion in others if anything exists having moved from potency to act in generation and corruption, or non-being to being in *creatio ex nihilo*.[477]

If 'b' necessarily cannot move itself, the antecedent again necessarily implies the consequent that of the existence of a first mover 'a' that

---

477 Aquinas likewise explains: "Whatever belongs to a thing is either caused by the principles of its nature (as the capacity for laughter in man) or comes to it from an extrinsic principle (as light in the air from the influence of the sun). Now being itself cannot be caused by the form or quiddity of a thing (by 'caused' I mean by an efficient cause), because that thing would then be its own cause and it would bring itself into being, which is impossible. It follows that everything whose being is distinct from its nature must have being from another. And because everything that exists through another is reduced to that which exists through itself as to its first cause, there must be a reality that is the cause of being for all other things, because it is pure being. If this were not so, we would go on to infinity in causes, for everything that is not pure being has a cause of its being, as has been said. It is evident, then, that an intelligence is form and being, and that it holds its being from the first being, which is being in all its purity; and this is the first cause, or God" (*On Being and Essence* 4.7). Mauer notes "See *Contra Gentiles*, I, 22, § 6. The form is not the efficient, but the formal cause, of being... God is pure being (*esse tantum*), being itself (*ipsum esse*), subsistent being (*esse subsistens*). He is not a being (*ens*), that is to say, a thing that participates in being in a finite way. See In Librum de Causis, 6; ed. H. D. Saffrey, p. 47; *Summa Theologiae*, I, 44, 1; I, 13, 11; *De Substantiis Separatis*, 8; ed. F. Lescoe, § 42, p. 79. See E. Gilson, *The Christian Philosophy of St. Thomas Aquinas* (New York, 1955), pp. 84-95; E. Gilson, *Elements of Christian Philosophy* (New York, 1960), pp. 124-133." Citation and notes from *Aquinas on Being and Essence*, trans. by A. A. Maurer (Toronto: PIMS, 2nd ed., 1968).

can move itself and that must necessarily move 'b' for 'b' to move. For example, by analogy, given that objects 'b' and 'c' are inanimate objects and 'a' is an animate object with motion in itself and capable of moving another, if 'c' were moved by 'b', if there were no first mover 'a' moved in itself and that can move another then 'b' and 'c' would remain stationary and not be moved being inanimate stationary objects that cannot move themselves nor move another. An infinite series is necessarily impossible because motion itself implies a first mover that is moved in itself and that can move another since it is impossible for a being to be at the same time and in the same respect in potentiality and in actuality to being and must be acted upon by something in act prior to coming to be and prior to moving something from potentiality to act. A similar argument holds true of a series of efficient causes as will be seen in the *Secunda Via*.

Since it is necessarily the case that a first mover must exist for a series of secondary movers to move from potency to act, <u>secondary movers can only (necessarily) effect motion if they are moved by a first mover</u>. It is this first mover of being (or motion) from potency to act from which the initial motion from potency to act obtains. It is this first mover that everyone holds upon rational reflection *or naturaliter nota* to be God for without a first mover nothing could have moved from potential being to actual being nor non-being to being since nothing of itself can move itself from potency to act nor from non-being to being.

Although Aquinas has in mind movement from potency to act in the *Prima Via*, the argument from non-being to being cannot be far from Aquinas's intent. In either case, the extrinsic principle or Prime Mover is implied in the motion between two extremes, either potency and act or non-being and being since a thing is reduced from potency to act or from non-being to being by some agent in act and this is necessarily the case and therefore the necessary existence of a Prime Mover has been demonstrated.

From necessary modal conditionals that are universal and necessary, one arrives at a first mover *a priori*. Further, since the argument is not only universal and necessary but is arrived at from the senses for it "is certain, and obvious to the senses, that in this world some things are moved," the argument is likewise *synthetic*. The conditional necessity is abstracted from the nature of things, and thus the necessity itself is *synthetic* while being necessary and universal since natural necessity is abstracted from sensible things as universal. Thus one can attain a synthetic *a priori* natural knowledge that the First Mover exists and by *naturaliter nota* this First Mover is so considered to be God.

Naturaliter nota is a Scholastic–Latin phrase meaning **"known by nature"** or **"naturally known."** In philosophy and theology, especially in **Aquinas and the medieval scholastics,** it refers to truths that the human intellect grasps **immediately and universally without requiring demonstration or experience (*naturaliter nota quoad no*).** These are self-evident principles—things that are so basic to rational thought that they do not need to be proved.

In Aquinas: naturaliter nota often designates the **first principles of knowledge**—such as "the whole is greater than the part" or "it is impossible for something to be and not be at the same time and in the same respect" (principle of non-contradiction). They do not require demonstration and are called ***naturaliter nota quoad no*s.**

Relation to Aristotle: Following Aristotle's ***Posterior Analytics,*** Aquinas holds that all demonstration presupposes first principles that are not themselves demonstrated but are **naturaliter nota.**

Theology: Some truths about God (e.g., His existence, His unity) are said to be *naturaliter nota* in the sense that they can be known through natural reason—though Aquinas distinguishes between truths that are **self-evident in themselves** (e.g., "God exists" as seen from God's essence) and those that are only **self-evident to us** after reasoning (via the Five Ways).

Category	Meaning	Example	Notes
Naturaliter nota in se ("self-evident in itself")	A proposition is true by its very terms; the predicate is contained in the subject, whether we realize it or not.	"God exists" (Deus est) — since God's essence is existence itself (ipsum esse subsistens).	True in itself, but not immediately evident to us, because we cannot see God's essence directly.
Naturaliter nota quoad nos ("self-evident to us")	A proposition that is obviously true as soon as its terms are understood.	"The whole is greater than the part."	Requires no proof; grasped immediately by natural reason.
Not naturaliter nota (to us)	A truth that requires reasoning, demonstration, or experience to be known.	God's existence proved by the Five Ways.	Although "God exists" is in se self-evident, for us it must be demonstrated from effects (efficient causality, necessity, motion, etc.).

Naturaliter nota = truths self-evident to reason, known by the natural light of the intellect without need for demonstration. Some things are **self-evident in reality** (because of their essence). Others are **self-evident to human reason** (once the terms are understood). And some, though true in themselves, require **demonstration to us** because we don't grasp their essence directly. This is the case when discussing God and God's existence.

Although we know God's existence using natural reason, proofs are required to demonstrate the existence of God. Many proofs can be given and Part II of this two-part series will outline various proofs for God's existence from nature, from Bayesian probability, from modal logic and various other demonstrations including demonstrations grounded in empirical science.

The Secunda Via: The Argument from Efficient Causality

Aquinas's argument for God's existence from efficient causality in *SCG* I, 13.33 rests upon demonstrating that there cannot be an infinite number of efficient causes in a series thus indicating the need for a first efficient cause. Once demonstrated, Aquinas can argue that since there cannot be an infinite regress in efficient causality there must be a first efficient cause. Aristotle had illustrated the absurdity of an infinite regress using two different types of regresses. The first is that of an infinite series in the case of efficient causality and the other that of an infinite kind or genus in material causality. In the latter case, one genus cannot proceed from another genus *ad infinitum*, e.g., using Aristotle's dated illustration, flesh from earth, earth from air, air from fire, and so on without stopping. Or, if one takes a more contemporary illustration based on modern taxonomy one might find Homo Sapiens from Mammalia, Mammalia from Animalia *ad infinitum*.[478] In the former case, one within a series cannot proceed from another within

---

478 It should be noted that in contemporary taxonomy "the only taxon that has a clear biological identity is the species. Members of a species share a common gene pool because they breed with one another. So members of a species form a very real biological unit… The taxa above the level of species, however, do not have a clear biological identity. This is because taxonomists, or scientists who classify organisms, draw the lines between one genus and another and between one family and another" according to what they perceive to be biologically shared or what might appear to be common characteristics. For example, members of the cat family (Felidae) are grouped under different genera. The tiger is under the genus Panthera, the ocelot, the puma, and the house cat all fall under the genus Felis, and the cheetah belongs to a third genus, Acinonyx. If two species share many features but are separate biological units, they are classified as different species within the same genus. The species within a common genus share common characteristics for example the Felis domesticus or the common house cat falls under the genus Felis along with the mountain lion (Felis concolor) since they seem to share various characteristics such as teeth, feet, and claws. Groups of genera that seem to share many common characteristics are placed in larger units for classification called families. All of the various genera cat like animals including Panthera, Felis, and Acinonyx fall under the family Felidae. Several families make up an order for example the cat family (Felidae) and dog family (Canidae) fall under Carnivora or carnivores, or meat-eaters. Orders are then grouped into classes for example the order Carnivora share with Primates such as humans common characteristics of being warmblooded, have body hair, and produce milk for their young. As such both Carnivora and Primate orders are classified together under the class Mammalia.

Finally the classification developed by Linnaeus groups common classes under phylum and common phyla into one of two taxa called kingdoms either that of the Animalia or Plantae kingdom. The eighteenth century taxa developed by Linnaeus has been revised to include microoganisms and instead of two kingdoms there are at least six today and not all agree on the various groupings nor upon what constitutes common characteristics for one group or another. Any such classification is simply a useful tool for studying organism; however, the mistake, is to assume that the interconnected evolutionary tree or any taxonomy existing as a being of reason does in fact exist in *res naturae* or *entia naturae* based solely upon similar characteristics. Although one might be able to associate one genus with another, in Aristotle's example we find a similar mistake as the one described above in making a logical association a actual association between one genus and another as in flesh from earth, earth from air, air from fire. However, Aristotle's point is that given any association between various genres one cannot go on to infinity from one genus to another in an ordered series of efficient causes. (Citations from Kenneth R. Miller and Joseph Levine, *Biology* (New Jersey: Prentice Hall, 1998), 320-327.

that same series *ad infinitum*, e.g., a man sets aside his clothing because the air is warm, the air is warm because of the sun, the sun is hot because it generates heat, it generates heat by being moved by some other cause, that cause generates the generation of heat by something else, and so on to infinity.[479]

Aquinas with Aristotle argues that if one were to suggest an infinite series of causes as given above, no cause would be first since an infinite series can have no first cause since there could be no causes within a given series without the first cause since no cause can *per se* be the cause of itself otherwise the effect would be prior its cause within an ordered series of efficient causes. If there were no first cause, this would mean there could be no intermediary causes nor any effects. Since we clearly see this is not the case since there are many causes and effects, it is impossible to hold to the position that there is an infinite series of efficient causes.

Again, if the first cause was removed, all the intermediary causes would likewise be removed in an ordered series of efficient causes since it would be the case that there would be no first cause to cause the intermediaries. This is impossible since there are in fact intermediary causes, and therefore there must have been a first efficient cause and not an infinite regress in causes.[480] With this background, the specific argument used by Aquinas in the *Secunda Via* is as follows:

---

479 Aristotle explains: "Evidently there is a first principle, and the causes of things are neither an infinite series nor infinitely various in kind. For, on the one hand, one thing cannot proceed from another, as from matter, ad infinitum, e.g. flesh from earth, earth from air, air from fire, and so on without stopping; nor on the other hand can the efficient causes form an endless series, man for instance being acted on by air, air by the sun, the sun by Strife, and so on without limit" *Meta.* II, a (994a2-5). Also see *ST* I, q46 obj. 7 and reply 7.

480 Aquinas explains: "The Philosopher proceeds in a different way in 2 *Metaph.* Ia, 2 (994a 1) to show that it is impossible to proceed to infinity in efficient causes, and that we must come to one first cause, and this we call God. This is how he proceeds. In all efficient causes following in order, the first is the cause of the intermediate cause, and the intermediate is the cause of the ultimate, whether the intermediate be one or several. Now if the cause be removed, that which it causes is removed. Therefore if we remove the first the intermediate cannot be a cause. But if we go on to infinity in efficient causes, no cause will be first. Therefore all the others which are intermediate will be removed. Now this is clearly false. Therefore we must suppose the existence of a first efficient cause: and this is God" (*SCG* I, 13).

The second way is from the nature of the efficient cause. In the world of sense we find there is an order of efficient causes. There is no case known (neither is it, indeed, possible) in which a thing is found to be the efficient cause of itself; for so it would be prior to itself, which is impossible. Now in efficient causes it is not possible to go on to infinity, because in all efficient causes following in order, the first is the cause of the intermediate cause, and the intermediate is the cause of the ultimate cause, whether the intermediate cause be several, or only one. Now to take away the cause is to take away the effect. Therefore, if there be no first cause among efficient causes, there will be no ultimate, nor any intermediate cause. But if in efficient causes it is possible to go on to infinity, there will be no first efficient cause, neither will there be an ultimate effect, nor any intermediate efficient causes; all of which is plainly false. Therefore it is necessary to admit a first efficient cause, to which everyone gives the name of God.[481]

Aquinas in the above argument as in the *Prima Via* begins from sensible knowledge and that which is self-evident. In the case of sensible things, there is an order of efficient causes where one necessarily cannot find a case where there is an efficient cause that is a cause of itself in a given order of efficient causes. Aquinas reasons that if something were an efficient cause of itself, it would be prior to itself and this is impossible. Hence, the contrary must necessarily be the case: if an efficient cause cannot be the cause of itself, an efficient cause necessarily cannot be prior to itself.

Aquinas argues that nothing can be the efficient cause of itself. If something were its own efficient cause (p), it would have to be prior to itself in the order of efficient causes (q). But self-priority is impossible in that order (priority is irreflexive and act precedes potency). Therefore, self-causation is metaphysically impossible.

---

481 *ST* I, q. 2, a. 3.

Let p(x) = "x is an efficient cause of itself" and q(x) = "x is prior to itself in the efficient-causal order."

Aquinas' necessities:

(A1) □ ∀x (p(x) → q(x))   — necessarily, if self-causation held, self-priority would follow.

(A2) □ ∀x (¬q(x))   — necessarily, no thing is prior to itself (irreflexive priority).

Valid inference in normal modal logic K:

From (A1) and (A2): ∀x □ ¬p(x) (equivalently: ∀x ¬◇p(x)).

Propositional core:

□(p → q), □¬q ⊢ □¬p (equivalently ¬◇p).

The Ad Absurdum explanation: assume for contradiction ◇p (self-causation is possible). From □(p→q), by normality, ◇p implies ◇q. But □¬q entails ¬◇q. Contradiction. Therefore ¬◇p, hence □¬p. Quantified: for any a, □(p(a)→q(a)) and □¬q(a) yield □¬p(a); since a was arbitrary, ∀x □¬p(x).

The argument from efficient causality is valid independent of any metaphysical ground and it is de re.

We can also frame this in terms of causality and causal relations.

Let C(x,y) = "x efficiently causes y" and x ≺ y = "x is prior to y (efficient-causal order)." Then:

(AP1) □ ∀x∀y (C(x,y) → x ≺ y)   — causes are prior to effects (act precedes potency).

(AP2) □ ∀x ¬(x ≺ x)   — priority is irreflexive.

From (AP1) and (AP2): □ ∀x ¬C(x,x). Identifying p(x) with C(x,x) and q(x) with x ≺ x recovers (A1)–(A2). The figures below provide the proof both in predicate logic and in Kripke frame intuition.

Figure 1. Proof-flow

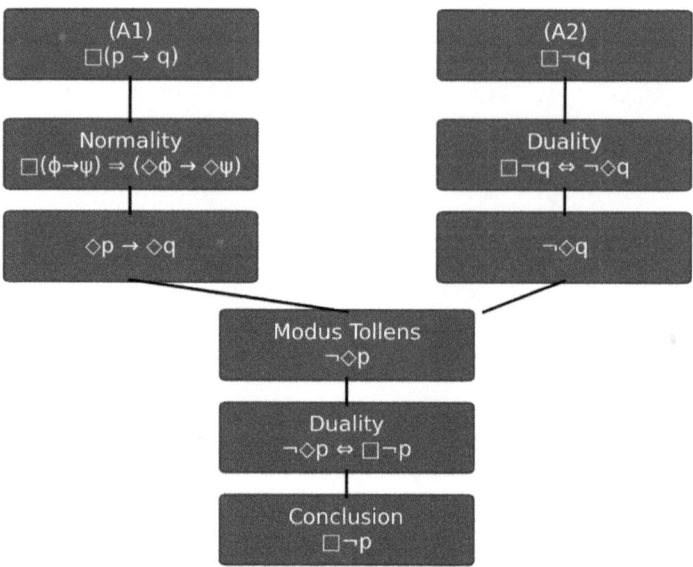

From □(p→q) and □¬q derive □¬p (valid in K)

Explanation:

A1 and A2 are the two modal premises fixed for Aquinas' step:

- **A1: □ ∀x (p(x) → q(x))**

  Reading: Necessarily, for every x, **if** x were the efficient cause of itself (p), **then** x would be prior to itself (q).

- **A2: □ ∀x (¬q(x))**

  Reading: Necessarily, for every x, it is **not** the case that x is prior to itself (self-priority is impossible).

(With p(x)p(x)p(x) = "x is an efficient cause of itself," q(x)q(x)q(x) = "x is prior to itself in the efficient-causal order.")

If we use causal-relation metaphysical representation: let C(x,y) C(x,y)C(x,y) = "x efficiently causes y," and x≺yx\prec yx≺y = "x is causally prior to y." Then:

- **A1** ≡ □ ∀x∀y **(C(x,y)** → **x≺y)** (causes are prior to effects), which yields the specific p(x)→q(x)p(x)→q(x)p(x)→q(x) when y=xy=xy=x.

- **A2** ≡ □ ∀x ¬**(x≺x)** (causal priority is irreflexive).

Setup:

Let:

p(x) := "x is an efficient cause of itself."

q(x) := "x is prior to itself (in efficient-causal order)."

We assume a normal modal system (at least K) with □ = necessity, ◇ = possibility, ∀ = universal quantifier.

Aquinas' Two Necessities:

(A1) □ ∀x (p(x) → q(x)) — Self-causation would entail self-priority, necessarily.

(A2) □ ∀x (¬q(x)) — Self-priority is necessarily impossible.

Valid Inference (in K):

From (A1) and (A2) it follows:

∀x □ ¬p(x) (equivalently: ∀x ¬◇p(x))

Proof Sketch:

Using a fixed arbitrary a. Then:

1. □(p(a) → q(a)) (from A1)
2. □¬q(a) (from A2)
3. From 1 and normality: □(p(a) → q(a)) ⊢ (◇p(a) → ◇q(a)).
4. From 2: ¬◇q(a).

5. By modus tollens on 3 and 4: $\neg\Diamond p(a)$.

6. Hence $\Box\neg p(a)$. Since a was arbitrary, $\forall x\ \Box\neg p(x)$.

Conclusion:

Aquinas' reasoning can be captured in modern modal logic:

$$\Box(p \rightarrow q),\ \Box\neg q \vdash \Box\neg p$$

Thus, self-causation is impossible, of necessity, since it would entail a contradiction in the order of act and potency. This also has significant implications for abiogenesis and the causality of existence.

Given that 'something were an efficient cause of itself' is (p) and 'something would be prior to itself' is (q) where p implies q, then '~ possibly p' implies 'necessarily ~ q'.

The *ad absurdum* argument amounts to the following rule of inference:

- **P(x)** = "x is an efficient cause of itself." This asserts *self-causation* — that the same being is both the efficient cause and the effect of its own existence or motion.

- **Q(x)** = "x is prior to itself in the order of efficient causality (act before potency)." This describes the necessary *priority relation* in efficient causation: a cause must be in act before that which it causes (which is in potency).

Aquinas's reasoning is then:

$$N\forall x\ (P(x) \supset Q(x)),\ N\forall x\ {\sim}Q(x) \supset N\forall x\ {\sim}P(x)$$

Let's parse it piece by piece:

- **N** = necessity operator ($\Box$).

  It signifies that the statement holds necessarily, not just contingently.

- $\forall x\ (P(x) \supset Q(x))$

  Means: for all x, if x is the efficient cause of itself, then x is prior to itself in the causal order.

(The definition of efficient causality entails such priority.)

- $N\forall x \ (P(x) \supset Q(x))$

  Necessarily, if anything were the cause of itself, it would have to be prior to itself.

  (This is an analytic truth about the concept of efficient causality.)

- $\forall x \ \sim Q(x)$

  Nothing can be prior to itself in the order of efficient causality.

  (That's metaphysically impossible.)

- $N\forall x \ \sim Q(x)$

  Necessarily, nothing can be prior to itself.

  (This impossibility is not contingent but metaphysically necessary.)

- $N\forall x \ \sim Q(x) \supset N\forall x \ \sim P(x)$

  Therefore, necessarily, if nothing can be prior to itself, then nothing can be its own efficient cause.

1. It is necessarily true that if something were its own efficient cause, it would have to be prior to itself in the order of efficient causality.

2. But it is necessarily true that nothing can be prior to itself.

3. Therefore, it is necessarily true that nothing can be its own efficient cause.

This formalization captures Aquinas's *rejection of self-causation* in modal terms:

- It is **logically impossible** (not merely empirically false) for something to cause itself.

- Causal priority entails *ontological asymmetry*: cause in act → effect in potency.

- Therefore, there must exist a *first efficient cause* not dependent on prior causes.

That premise — "nothing can cause itself" — becomes the foundation for the regress argument that leads to God as ipsum esse subsistens, the First Cause.

'N' is shorthand for 'necessarily'. The material implication $\supset$ indicates 'that where the antecedent is true the consequent follows' and the negation operator $\sim$ indicates 'it is not the case that'. The universal quantifier $\forall$ indicates universal necessity such as all things, everything, anything.[482] Potentiality cannot both precede and succeed act at the same time and in the same respect. Likewise, a cause cannot both be and not be the cause of itself.

Since it is the case in the order of efficient causes that an efficient cause cannot be its own cause and act necessarily precedes potentiality, it is impossible for an efficient cause to be prior to itself and thus the consequent necessarily follows from the antecedent condition.

The contrary also holds, since an efficient cause necessarily cannot be prior to itself, it is necessarily the case that an efficient cause cannot be the cause of its own self. Therefore '$\sim$ possibly p' implies 'necessarily $\sim q$' such that if it is not possible for something to be the efficient cause of itself then it is necessarily not the case that the efficient cause will be prior to itself. In a sense since the consequent is implied in the antecedent, it is analytic *a priori* as well as synthetic *a priori*.

Aquinas's assertion above is not expressed in a conditional proposition given by an 'if then' clause but expressed as absolute necessity: "There is no case known (neither is it, indeed, possible) in which a thing is

---

482 The four Aristotelian forms in categorical syllogisms are the following:

All p's are q's. $\forall x \, (p(x) \supset q(x))$
Some p's are q's $\exists x \, (p(x) \wedge q(x))$
No p's are q's $\forall x \, (p(x) \supset \sim q(x))$
Some p's are not q's $\exists x \, (p(x) \wedge \sim q(x))$

Adopted from Jon Barwise and John Etchemendy, *Language Proof and Logic*, (Stanford: CSLI Publications, 1999, 2000, 2002), 241. The use of modal sentences in the composite *(in sensu composito)* and divided sense *(in sensu diviso)* and particularly the use of de re modality as it refers to things the way they are as opposed to *de dicto* modality associated with the truth of what is said by the sentence see Henrik Lagerhund. According to Henrik Lagerhund, "Virtually all medievals held that Aristotle's modal syllogistic was a theory for *de re* modal sentences" and this seems to suggest that for virtually all medievals there was no disconnect between the way things are in modal logic and the universality and necessity found in judgment or reason (cf. Henrik Lagerhund, "Section 4 Peter Abelard" of "Medieval Theories of the Syllogism" in the *Stanford Encyclopedia*: http://plato.stanford.edu/entries/medieval-syllogism/ (February 19, 2005). Also see H. Lagerliund, *Modal Syllogistics in the Middle Ages*, (Leiden: Brill, 2000) and Simo Knuuttila, "Medieval Theories of Modality" in the Stanford Encyclopedia: http://plato.stanford.edu/entries/modality-medieval/ (February 19, 2005).

found to be the efficient cause of itself; for so it would be prior to itself, which is impossible."[483] The argument is not only self-evident from experience taken inductively, but also the argument is *a priori* as apprehended from experience and expressed by a modal conditional where the consequent is implied in the antecedent. However, it is not a tautology since the consequent is not identical to the antecedent. Having demonstrated that an efficient cause cannot be the cause of itself, Aquinas proceeds with the argument. Aquinas argues that it is *impossible* to go on to infinity among efficient causes, and what is implied is that there *necessarily* must be a first efficient cause.

Aquinas argues that in every case of ordered efficient causes, since an efficient cause cannot cause itself as demonstrated above, a first cause must necessarily be the cause of the intermediate efficient cause and so forth. It makes no difference on the number of causes since if one removes one cause in a series of only one efficient cause, or remove any number of intermediate causes or the first cause, the effect is removed in an ordered series of causes. Hence if one removed the first cause, there would be no last or intermediate since in a series that depends upon the first efficient cause for the existence of the second efficient cause and the second for the existence of the third and so forth. If the first efficient cause is removed, then the second is removed. If the second, then the third and so forth.

Likewise, if one removed the first cause by having an infinite ordered series of causes, it would necessarily be the case that there could be no last cause nor any intermediate causes. Hence, there can be no infinite series of causes at all within an ordered series of causes because without a first cause there can be no second cause and without a second

---

483 The 'if then' clause or the material operator inadequately expresses Aquinas's absolute necessity. Aquinas indicates it is strictly impossible (necessarily not the case) for q to obtain *since* it is not possible for p to obtain. One might be able to state the material implication in 'while ~Mp then N~q', but this again fails to express the absolute necessity being conveyed since it will never be the case that an efficient cause will be the cause of itself and a "while clause" does not adequately express the proper material implication since a "while clause" like an "if clause" lends itself to the possibility of the antecedent obtaining under some condition.

then no third etc. Hence an infinite ordered series of efficient causes is not possible. Since there cannot be an infinite series of efficient causes because one cannot have an ordered series of causes without a first efficient cause, one is necessarily left with a first efficient cause of a finite ordered series of efficient causes. Consequently, if and only if there is a finite ordered series of causes, there must necessarily be a first efficient cause of a finite ordered series of efficient causes.

It is this first efficient cause, in a finite ordered series of efficient causes, that everyone considers to be God. This first efficient cause, by which all things exist: is taken *naturaliter nota* or a priori analytic to be God. The reason is that without a first efficient cause, there can necessarily be no efficient cause within a series, or within a kind, or a genus. They cannot have been caused since nothing of itself can cause itself nor can an efficient cause be prior to itself moving from potency to act.[484] Nothing can move from potency to act except by something in act and it is this first pure act which is the uncaused cause that everyone considers *naturaliter nota* to be God. An example might be the notion of abiogenesis. For some y to come to exist, it must necessarily have some x that exists for y to come into existence. The building blocks of life must exist, and yet they could not have caused themselves and thus we have a material infinite regress what made x if x is necessarily for y to exist.

Aquinas in his earlier work the *Summa Contra Gentiles* went beyond his introductory text in the *Summa Theologiae* to demonstrate that there must be a cause to which it belongs to give existence to all things and this cause is necessarily God.

---

484 Although an effect may be simultaneous with its cause or multiple causes may be a cause of an effect, the effect itself cannot be its own cause since the effect cannot be prior to its cause when speaking of an ordered series of efficient causes such as A B C D where B follows A at time t1, C follows B at time t2, etc. As noted in the following footnote, *from the efficient cause it follows that the effect is necessarily, when it is necessary for the agent to act, for it is through the agent's action that the effect depends on the efficient cause* (SCG II, 31). In a series of efficient causes it is necessary for a first efficient cause to act for the finite series to exist, for it is though the agent's action that the effect depends on the efficient cause. Likewise an infinite series is impossible for without a first efficient cause there cannot be a series of causes since it is through the first efficient agent's action that the effect or any intermediate or secondary efficient cause depends and hence there can be no infinite series of efficient causes and effects.

Since there must necessarily be a first efficient cause, for there to be any intermediate causes, or a last cause. Because a cause necessarily cannot have caused itself, it is necessary for a first efficient cause to exist. Being cannot come from non-being and an effect is not prior to its cause, therefore it is necessarily the case that there is a first efficient cause that is not caused by another.[485] And this cause to exist without coming into existence, must necessarily be self-subsistent and incorruptible and not contingent being and therefore not material. Matter is necessarily contingent and corruptible. This first efficient cause by which all things exist is taken *naturaliter nota* to be God for without a first efficient cause nothing could have been caused since nothing of itself can cause itself moving from potency to act.[486]

When arguing from the communication of *esse*, Aquinas argues that an effect must be proportionate to its cause but since all things

485 Aquinas explains: "That which is the first of beings, must needs be the cause of the things that are: for if they were not caused they would not be set in order thereby, as we have already proved. Now between act and potentiality there is this order, that, although in the one and same thing which is sometimes in potentiality and sometimes in act, potentiality precedes act in point of time, whereas act precedes by nature; nevertheless, speaking simply, act must needs precede potentiality, which is evidenced by the fact that potentiality is not reduced to act save by a being in act. But matter is a being in potentiality. Therefore God Who is pure act must needs be simply prior to matter, and consequently the cause thereof. Therefore matter is not necessarily presupposed for His action.

Again. Primary matter is in some way, for it is a being in potentiality. Now God is the cause of all things that are, as we have proved [SCG XV]. Therefore God is the cause of primary matter: to which nothing is pre-existent. Therefore the divine action needs no pre-existing nature. Divine Scripture confirms this truth, saying (Gen. i. 1): In the beginning God created heaven and earth. For to create is nothing else than to bring something into being without preadjacent matter. Hereby is refuted the error of the ancient philosophers who asserted that matter has no cause whatever, because they observed that in the actions of particular agents something is always preadjacent to action: whence they drew the opinion common to all that from nothing naught is made [Physics I, 4.2]. This is true in particular agents. But they had not yet arrived at the knowledge of the universal agent, which is the active cause of all being, and of necessity presupposes nothing for His action" (SCG II, 16).

486 Also a similar demonstration can be drawn from SCG II, 20, 16 explaining that since it is impossible for a body to act save by contact and since such contact is of one thing in relation to another, a bodily agent cannot act by creation because there is nothing that pre-exists for that bodily agent to be in contact with since matter is a being in potentiality and nothing in potentiality is reduced to act save by another in act. "Again since mover and moved, maker and made must be together as proved in Ari: Physics VII, 2," and since "a bodily agent cannot be present to its effect except by contact, whereby the extremes of contiguous things come together as demonstrated in Ari: Physics V, 3.8," there must necessarily either be nothing in existence if there is no pre-existent matter in which the mover is in contact or there must be a pure act who is prior to matter from which *being* itself was caused. Since it is impossible for nothing to be in existence since we know that being exists, either there was pre-existence matter or there must be a pure act, since it has been demonstrated that there cannot be pre-existent matter there necessarily can only be a pure act that is prior to matter from which being itself is caused.

have existence that goes beyond this proportionality, it means that something goes beyond this or that things immediate cause in giving things their act of existence.[487] Therefore there must be a cause to which

487 Aquinas provides a number of demonstrable proofs that God is the first being and first cause of all being in *SCG* II, 15 and *SCG* II, 16. Aquinas also provides arguments that do not "conclude of absolute necessity, but that are not devoid of probability" against arguments for an eternally existing universe in *SCG* II, 38, and in *SCG* I, 3 and *SCG* II, 31 Aquinas indicates that any argument *for eternal existence* will likewise be an argument from probability. Aquinas argues that matter in and of itself cannot be the cause of the universe since matter is in potentiality and not a principle of activity (*SCG* I, 17). Aquinas notes that Aristotle had demonstrated that efficient and material causes do not coincide (*SCG* I, 17; *Phys.* II, 7.3). Aquinas in arguing that God brought things into being out of nothing states: "between act and potentiality there is this order, that, although in the one and same thing which is sometimes in potentiality and sometimes in act, potentiality precedes act in point of time, whereas act precedes by nature; nevertheless, speaking simply, act must needs precede potentiality, which is evidenced by the fact that potentiality is not reduced to act save by a being in act. But matter is a being in potentiality. Therefore God Who is pure act must needs be simply prior to matter, and consequently the cause thereof" (*SCG* II, 16).

Aquinas thus concludes that since matter is in potentiality to being acted upon by some prior agent, matter itself cannot be that agent that acts upon itself since being cannot come from non-being independent of an agent that brings something from nothing. Nor can that which is in potentiality to being be moved to being except by some being in act. Aquinas addresses both a third agent as well as eternal matter. Hence God who is pure and eternal act is the efficient causal agent that acts on matter (*SCG* II, 16). Therefore God is the cause of matter, and since God is the cause of all things (*SCG* II, 15) divine action generates matter and has no need for a pre-existing nature or matter in causing the universe to be (*SCG* II, 16). Aquinas explains, "hereby is refuted the error of ancient philosophers who asserted that matter has no cause whatever, because they observed that in the actions of particular agents something is always prejacent to action: when they drew the opinion common to all that from nothing naught is made. This is true in particular agents (such as matter). But they had not yet arrived at the knowledge of the universal agent, which is the active cause of all being, and of necessity presupposes nothing for His effects" (*SCG* II, 16).

Aquinas argues that arguments for a beginning, like eternal existence, can only be argued for by probable arguments. That the universe had a beginning is taken from revelation as an article of faith: "by faith alone do we hold, and by no demonstration can it be proved, that the world did not always exist, as was said above of the mystery of the Trinity (*ST* I, q32, a1). The reason for this is that the newness of the world cannot be demonstrated on the part of the world itself. For the principle of demonstration is the essence of a thing. Now everything according to its species is abstracted from 'here' and 'now'; whence it is said that the universals are everywhere and always. Hence it cannot be demonstrated that men, or heaven, or a stone were not always. Likewise neither can it be demonstrated on the part of the efficient cause, which acts by will. For the will of God cannot be investigated by reason, … but the divine will can be manifested by revelation, on which faith rests. Hence that the world began to exist is an object of faith, but not of demonstration or science" (*ST* I, q. 46, a. 2).

Interestingly Aquinas qualifies his position in a reply to objection 1 in the same question (*ST* I, q. 46, a. 2 reply obj. 1). The objection to creation being an article of faith states that a) everything made has a beginning as to its duration and b) that it can be proved demonstrably that God is the efficient cause of the world therefore it can be demonstrably proved that the world began. However, Aquinas replies to this mistaken objection by explaining that although it is true that arguments that hold that the world was not from God can be refuted by proofs that are cogent, it cannot be demonstrated that creation was not instantaneous and not successive as an act of God's creation. Therefore God in being the active cause of the world does not have to be necessarily prior in duration to the world since creation is not a successive change (*ST* I, q45, a2). It seems to follow from this that even if one were able to demonstrate

it belongs to give existence to all things. Since it has been demonstrated above that God is the first cause, it follows therefore that everything that exists, are from God from whom they derive their act of existence.[488]

From necessary modal conditionals that are both universal and necessary, one arrives at a first efficient cause *a priori*. Furthermore, since the argument is not only universal and necessary, but also arrived at from the senses for "we find that among sensible things there is an ordering of efficient causes, and yet we do not find—nor is it possible to find [necessarily not the case or impossible]—anything that is an efficient cause of its own self," the argument is likewise *synthetic due to extension and a posteriori*. The conditional necessity is abstracted from the nature of things, and thus the necessity itself is *synthetic* while being necessary and universal since natural necessity is abstracted from

from the natural sciences that the universe had a beginning, from logical necessity one cannot demonstrate that the universe had a beginning. Hence, God could have eternally generated existence and logical necessity cannot demonstrate otherwise.

However, Aquinas would likely admit that even though logical necessity cannot demonstrate that the universe had a beginning, this would not be the case with natural necessity (necessity from the way things are). From natural necessity, if it were demonstrated from the natural sciences that the universe had a beginning necessarily, it would necessarily follow that the universe began. For Aquinas in the 13th century it could not have been foreseen that natural science might actually be able to demonstrate from nature the possibility of a beginning both from cosmology as well as in the geological record. Thus Aquinas maintained that it could not be proven that the universe could not have always existed although arguments that reject God as the cause could be refuted (see *SCG* II, 38). Because of this in *ST* 46, quoted above, Aquinas maintained: "The reason for this [that the universe did not always exist] is that the newness of the world cannot be demonstrated on the part of the world itself" (*ST* q. 46, a. 2). However, if natural necessity demonstrates a beginning then the universe cannot be eternal and this only leaves the possibility that either some third cause created the universe, or God himself created. Aquinas seems to successfully refute the notion of a third cause based upon ordered causality and the failure of chance to obtain in matters of natural necessity in *SCG* II, 15; *SCG* II, 16; Arist. *Phys.* II, 4.7, 8. Therefore, if the philosophers arguments are valid and sound, only God can be the creator of existence if nature itself indicates that the universe began.

488 Aquinas explains: "The order of causes must needs correspond to the order of effects, since effects are proportionate to their causes [*Phys.* II, 3.12]. Wherefore, as proper effects are reduced to their proper causes, so that which is common in proper effects must needs be reduced to some common cause: even so, above the particular causes of the generation of this or that thing, is the sun the universal cause of generation; and the king is the universal cause of government in his kingdom, above the wardens of the kingdom and of each city. Now being is common to all. Therefore above all causes there must be a cause to which it belongs to give being. But God is the first cause, as shown above. Therefore it follows that all things that are, are from God." *SCG* II, 15. Aquinas also argues that *esse* is communicated to existence because God himself is being as such: "That which is said to be essentially so and so is the cause of all that are so by participation: thus fire is the cause of all things ignited as such. Now God is being by His essence, because He is being itself: whereas everything else is being by participation: for there can be but one being that is its own being, as was proved in the First Book. Therefore God is the cause of being to all other things" (*SCG* II, 15).

sensible things as universal. Thus, one can attain a synthetic *a priori* natural knowledge that a First Efficient Cause necessarily exists and by *naturaliter nota* this First Efficient Cause is considered to be God.

The Tertia Via: The Argument from Necessity

Aquinas's third argument for God's existence in the *Summa Theologiae* I, q. 2, a. 3 argues from possibility and necessity.[489] The argument is given below:

> The third way is taken from possibility and necessity, and runs thus. We find in nature things that are possible to be and not to be, since they are found to be generated, and to corrupt, and consequently, they are possible to be and not to be. But it is impossible for these always to exist, for that which is possible not to be at some time is not. Therefore, if everything is possible not to be, then at one time there could have been nothing in existence. Now if this were true, even now there would be nothing in existence, because that which does not exist only begins to exist by something already

---

489 Also Aquinas writes: "Because if it be necessary for the universe of creatures, or any particular creature whatsoever, to be, it must have this necessity either of itself or from another. But it cannot have it of itself. For it was proved above (*SCG* II, 15) that every being must be from the first being. Now that which has being, not from itself, cannot possibly have necessity of being from itself; since what must necessarily be, cannot possibly not be; and consequently that which of itself has *necessary* being, has of itself the *impossibility* of not being; and therefore it follows that it is not a non-being; wherefore it is a being.

If, however, this necessity of a creature is from something else, *it must be* from a cause that is extrinsic; because whatever we may take that is within the creature, has being from another. Now an extrinsic cause is either efficient or final. From the efficient cause, however, it follows that the effect is *necessarily, when* it is *necessary* for the *agent to act, for it is through the agent's action that the effect depends on the efficient cause*. Accordingly if it is not necessary for the agent to act in order that the effect be produced, neither is it absolutely necessary for the effect to be. [However, as shown above if it is necessary for the agent to act in order that the effect be produced, it is absolutely necessary for the effect to be the agent's action since the effect depends on its efficient cause. Now God does not act of necessity in producing creatures, as we have proved above. Wherefore it is not absolutely necessary for the creature to be, as regards necessity dependent on the efficient cause" (*SCG* II, 31). The effect is necessary when it is necessary for an agent to act since it is through the agent's action that the effect depends on the efficient cause; however, since it is not necessary for God to act it is likewise not absolutely necessary for the creature to be, as regards necessity dependent on the efficient cause. Therefore Aquinas concludes in *SCG* II, 31 that existence is not necessary since God willed creation into existence and could have chosen otherwise, but since the universe does in fact exist, it has necessary existence in that it does in fact exist and is not simply probable but it is not absolutely necessary as regards necessity dependent on efficient causality.

existing. Therefore, if at one time nothing was in existence, it would have been impossible for anything to have begun to exist; and thus even now nothing would be in existence--- which is absurd. Therefore, not all beings are merely possible, but there must exist something the existence of which is necessary. But every necessary thing either has its necessity caused by another, or not. Now it is impossible to go on to infinity in necessary things which have their necessity caused by another, as has been already proved in regard to efficient causes. Therefore we cannot but postulate the existence of some being having of itself its own necessity, and not receiving it from another, but rather causing in others their necessity. This all men speak of as God.[490]

The argument moves from possible being to the existence of some being having of itself its own necessity not received from another, but rather causes in others their necessity. Similar to the argument from motion in the *Prima Via*, this implies possible and necessary existence where necessary existence is caused by something's own necessity, or the necessity is caused by another. Something has possible existence if it has the possibility of existing or not existing, and something has necessary existence if it does in fact exist.

Thomas argues that in nature we find things that have the possibility of existing or not existing, since they are generated or corrupted and therefore they have the possibility of existing or not existing depending upon if they are generated or if they decay. For example, a flower may bloom and then later decay. It is necessarily impossible for these always to exist since they are generated or corrupted and thus have the possibility of not existing and what does not always exist does not exist at some point otherwise it would not have the possibility of not existing. Therefore, if everything has the possibility of not existing due to decay, then at one time there *could have been* nothing in existence. Now if this possibility were actually the case or true, even now there

---

490 *ST* I, q. 2, a. 3.

would be nothing in existence, because that which does not exist only begins to exist by something already existing. Therefore, if it is possible that at one time nothing was in existence, then it is likewise possible that it would have been impossible for anything to have begun to exist; and if this possibility were actually the case even now nothing would be in existence---which is absurd.

Therefore, since it is necessarily the case that not all beings are merely possible since beings do in fact exist, there must exist something the existence of which is necessary since existing beings necessarily implies a necessary being. For necessary beings cannot exist apart from a being that necessarily exists of itself and that causes necessity in others. This follows from the fact that a) every necessary thing either has its necessity caused by another, or not, b) in a finite series of necessary beings, every necessary thing except the first has its necessity caused by another,[491] c) it is impossible to go on to infinity in necessary things which have their necessity caused by another, as has been already proved in regard to efficient causes,[492] and from the conclusion that follows from (a, b, and c) that we cannot but postulate the existence of

---

491 This is implied in a and c since a necessary thing cannot cause its own necessity if it is a secondary cause.
492 Aquinas argues that it is *impossible* to go on to infinity among efficient necessary causes, and what is implied is that there *necessarily* must be a first efficient necessary cause or mover. Aquinas argues that in every case of ordered efficient necessary causes, since an efficient necessary cause cannot cause itself, a first necessary cause must necessarily be the cause of the intermediate and so forth for any intermediate or secondary cause to exist. It makes no difference on the number of causes since if one removes one cause in a series of only one efficient necessary cause, or if any number of intermediate necessary causes are removed, or if the first efficient necessary cause is removed, the succeeding effect or the next efficient necessary cause in the series is removed. Hence if one removed the first cause, there would be no last or intermediate since in a series that depends upon the first efficient cause for the existence of the second efficient cause and the second for the existence of the third and so fourth, if the first efficient cause is removed then the second is removed and if the second then the third and so forth. This likewise applies to necessary causes or beings. Likewise, if one removed the first cause by having an infinite series of causes, again it would necessarily be the case that there could be no last cause nor any intermediate cause and thus no infinite series of causes at all because without a first cause there can be no second cause and without a second then no third etc. Hence one is either left with no infinite series of efficient necessary causes or one is left with a first efficient necessary cause or being of a finite series of efficient necessary causes. Since there cannot be an infinite series of efficient necessary causes because one cannot have a series of necessary causes without a first efficient necessary cause, one is necessarily left with a first efficient necessary cause of a finite series of efficient necessary causes. This first efficient necessary cause is absolutely necessary if one is to have a series of efficient necessary causes. This first efficient necessary cause is Aquinas's necessary being having of itself its own necessity, and not receiving it from another, but rather causing in others their necessity.

some being having of itself its own necessity, and not receiving it from another, but rather causing in others their necessity. This all men speak of as God. This can be formalized as follows:

Aquinas's Third Way: Formal Argument with Flow Commentary

Core Notation

$E(x)$: x exists

$N(x) := \Box E(x)$: x is necessary

$C(x) := \Diamond \neg E(x)$: x is contingent

$Dep(x,y)$: x depends on y

$Term(r)$: r is a terminal sustainer

$EssEqEx(x)$: essence of x = existence

Formal Proof with Aquinas's Flow

1. $\forall x \ (x = x)$ (LOI)

Flow: Identity: establishes determinate subjects. Aquinas begins by noting things have definite natures.

2. $\forall x \ \neg(E(x) \wedge \neg E(x))$ (PNC)

Flow: Non-Contradiction: secures that beings cannot both exist and not exist simultaneously.

3. $\forall x \ (E(x) \vee \neg E(x))$ (LEM)

Flow: Excluded Middle: beings either exist or do not. This allows us to talk coherently about possible non-being.

4. $C(x) := \Diamond \neg E(x), N(x) := \Box E(x)$

Flow: Definitions: contingency = possible non-existence; necessity = cannot not exist.

5. $A \rightarrow \Diamond A$ (RP)

Flow: Actuality implies possibility: what is actual is possible. Aquinas uses this to bridge observation and modal reasoning.

6. $\exists x\ \exists t1\ \exists t2\ (E(x,t1) \wedge \neg E(x,t2))$

Flow: Observation of change: beings come into and go out of existence. Aquinas appeals to experience of corruptible things.

7. Let $c,t1,t2$ witness (6). Then $E(c,t1)$ and $\neg E(c,t2)$.

Flow: Concrete example: some being c shows existence at one time and not at another.

8. From $\neg E(c,t2)$: $\neg E(c)$ at t2.

Flow: Therefore c fails to exist at some time.

9. From (5): $\neg E(c) \rightarrow \Diamond \neg E(c)$.

Flow: By RP, if c fails at some time, it is possible for c not to exist.

10. Hence $\Diamond \neg E(c)$.

Flow: Thus, c is contingent.

11. By def of C: $C(c)$.

Flow: So c is formally classified as contingent.

12. $\exists x\ C(x)$.

Flow: Conclusion of first block: contingent beings exist. Aquinas's starting point.

13. $\forall x\ (C(x) \rightarrow \exists y\ Dep(x,y))$ (PSR)

Flow: Sufficient Reason: contingent beings require a cause or sustainer. Aquinas: 'what can fail to be must be caused by another.'

14. $\forall x\ (C(x) \rightarrow \neg Dep(x,x))$ (Act–Potency)

Flow: No contingent can sustain itself. Aquinas rules out self-causation in sustaining existence.

15. $\forall x\ (C(x) \rightarrow \exists r\ (Chain(x,r) \wedge Term(r)))$ (No Regress)

Flow: Dependence chains terminate in a root. Aquinas denies infinite regress in per se causes of existence.

16. $\forall x\ (N(x) \rightarrow EssEqEx(x))$ (Essence=Existence)

Flow: In a necessary being, essence and existence are identical. Aquinas's metaphysical capstone.

17. From (12): pick a with C(a).

    Flow: We now examine one contingent being as representative.

18. From (15): ∃r (Chain(a,r) ∧ Term(r)). Choose r.

    Flow: This contingent traces back to a terminal sustainer r.

19. Term(r) ⇒ ¬∃y Dep(r,y).

    Flow: By definition, r depends on nothing else.

20. Suppose ◇¬E(r). Then C(r).

    Flow: Assume for contradiction that the root is contingent.

21. From (13): ∃y Dep(r,y).

    Flow: But contingents require sustainers by PSR.

22. Contradiction with (19).

    Flow: This contradicts r being terminal. Aquinas: the root cannot be contingent.

23. Hence ¬◇¬E(r). By duality, □E(r). So N(r).

    Flow: Therefore, the root sustainer is necessary. Aquinas: 'Therefore, something must exist of itself, necessary.'

24. From (16): N(r) ⇒ EssEqEx(r).

    Flow: In this necessary being, essence and existence coincide.

25. So N(r) ∧ EssEqEx(r).

    Flow: The root sustainer is necessary and self-subsistent.

26. ∃n (N(n) ∧ EssEqEx(n)).

    Flow: Final conclusion: there exists a necessary being whose essence is existence. Aquinas: 'and this all men call God.'

For a detailed explanation, let's walk through the notation, the argument, and proof embedded in the argument:

Full Notation & Operators with Explanation

$\forall x$ — Universal quantifier — "for all x."

$\exists x$ — Existential quantifier — "there exists some x."

a, b, c, r, t — Individual constants (stand for particular beings or times).

$E(x)$ — "x exists."

$E(x,t)$ — "x exists at time t." (represents change).

$Dep(x,y)$ — "x depends essentially (per se) on y."

$Chain(x,r)$ — "r is in the dependence chain of x."

$Term(r)$ — "r is a terminal sustainer (root, depends on nothing else)."

$EssEqEx(x)$ — "The essence of x is identical with existence."

$\Box A$ — Necessity — "A must be true."

$\Diamond A$ — Possibility — "A may be true."

Duality — $\Box A \equiv \neg\Diamond\neg A$.

$N(x)$ — $\Box E(x)$: x is necessary.

$C(x)$ — $\Diamond\neg E(x)$: x is contingent.

$\neg A$ — Negation — "not A."

$A \wedge B$ — Conjunction — "A and B."

$A \vee B$ — Disjunction — "A or B."

$A \rightarrow B$ — Implication — "if A, then B."

$A \leftrightarrow B$ — Biconditional — "if and only if."

(LOI) — $\forall x \ (x = x)$: Law of Identity.

(PNC) — $\forall x \ \neg(E(x) \wedge \neg E(x))$: Law of Non-Contradiction.

(LEM) — $\forall x \ (E(x) \vee \neg E(x))$: Law of Excluded Middle.

(PSR) — $\forall x \ (C(x) \rightarrow \exists y \ Dep(x,y))$: Principle of Sufficient Reason.

(Act–Potency) — $\forall x \ (C(x) \rightarrow \neg Dep(x,x))$: No contingent sustains itself.

(No Regress) — $\forall x \ (C(x) \rightarrow \exists r \ (Chain(x,r) \wedge Term(r)))$: Every chain of contingent dependence terminates in a root.

(Essence=Existence) — $\forall x \ (N(x) \rightarrow EssEqEx(x))$: If necessary, then essence = existence.

(RP) — $A \rightarrow \Diamond A$: Actuality implies possibility.

Observation — $\exists x \ \exists t1 \ \exists t2 \ (E(x,t1) \wedge \neg E(x,t2))$: There is some being that exists at one time and fails to exist at another (change).

Formal Argument and Proof:

1. $\forall x\ (x = x)$ (LOI) Law of Identity: establishes that things are determinate subjects of predication.

2. $\forall x\ \neg(E(x) \wedge \neg E(x))$ (PNC) Law of Non-Contradiction: existence and non-existence cannot both apply in the same respect.

3. $\forall x\ (E(x) \vee \neg E(x))$ (LEM) Law of Excluded Middle: for any being, either it exists or it does not.

4. $C(x) := \Diamond\neg E(x)$, $N(x) := \Box E(x)$ Definitions: contingency and necessity introduced using modal operators.

5. $A \rightarrow \Diamond A$ (RP) Actuality implies possibility: what is actual is at least possible.

6. $\exists x\ \exists t1\ \exists t2\ (E(x,t1) \wedge \neg E(x,t2))$ Observation: beings come into and pass out of existence (change). Intelligible due to LOI–PNC–LEM.

7. Let c,t1,t2 witness (6). Then $E(c,t1)$ and $\neg E(c,t2)$. Existential instantiation: a particular being c shows change.

8. From $\neg E(c,t2)$: $\neg E(c)$ at t2. Shows c is not necessary, since it fails at some time.

9. From (5) with A:=$\neg E(c)$: $\neg E(c) \rightarrow \Diamond\neg E(c)$. By RP, non-existence entails possible non-existence.

10. Hence $\Diamond\neg E(c)$. Therefore, c is contingent.

11. By def of C: $C(c)$. Restates contingency definition.

12. $\exists x\ C(x)$. Therefore, contingent beings exist.

13. $\forall x\ (C(x) \rightarrow \exists y\ Dep(x,y))$ (PSR) Every contingent being depends on another.

14. $\forall x\ (C(x) \rightarrow \neg Dep(x,x))$ (Act–Potency) No contingent sustains itself.

15. $\forall x\ (C(x) \rightarrow \exists r\ (Chain(x,r) \wedge Term(r)))$ (No Regress) Dependence chains of contingents terminate in a root.

16. $\forall x\ (N(x) \rightarrow EssEqEx(x))$ (Essence=Existence) A necessary being's essence is existence.

17. From (12): pick a with $C(a)$. Take a contingent being for analysis.

18. From (15): $\exists r\ (Chain(a,r) \wedge Term(r))$. Choose r. Every contingent leads to a root sustainer r.

19. $Term(r) \Rightarrow \neg\exists y\ Dep(r,y)$. Root sustainer depends on nothing further.

20. Suppose $\Diamond\neg E(r)$. Then $C(r)$. Assume for contradiction that the root is contingent.

21. From (13): ∃y Dep(r,y). By PSR, a contingent must depend on something else.

22. Contradiction with (19). This violates the definition of root sustainer.

23. Hence ¬◇¬E(r). By duality, □E(r). So N(r). Therefore r is necessary.

24. From (16): N(r) ⇒ EssEqEx(r). Necessary being entails essence = existence.

25. So N(r) ∧ EssEqEx(r). Root sustainer is necessary and self-subsistent.

26. ∃n (N(n) ∧ EssEqEx(n)). Therefore, a necessary being exists whose essence is existence — God.

From natural modal possibility and necessity Aquinas concludes that it must necessarily be the case that not all beings are merely possible otherwise necessarily nothing would even now be in existence developing an *ad absurdum* argument from modal conditional necessity, but since things do in fact exist, there must be something that necessarily exists.[493] Like the *Prima Via* and the *Secunda Via,* the *Tertia Via* argues from necessary modal conditionals that are both universal and necessary, the conclusion of the existence of a necessary being is therefore a *priori*. Further, since the argument is not only universal and necessary but is arrived at from the senses for "we find in nature

---

493 This again reflects the argument given in *SCG* II, 31: "Because if it be necessary for the universe of creatures, or any particular creature whatsoever, to be, it must have this necessity either of itself or from another. But it cannot have it of itself. For it was proved above (*SCG* II, 15) that every being must be from the first being. Now that which has being, not from itself, cannot possibly have necessity of being from itself: since what must necessarily be, cannot possibly not be; and consequently that which of itself has *necessary* being, has of itself the *impossibility* of not being; and therefore it follows that it is not a non-being; wherefore it is a being" (emphasis added). It must necessarily be the case that not all beings are merely possible otherwise nothing would even now be in existence since what is cannot not be, and therefore since things do in fact exist, there must be something that necessarily exists. If it is necessary for the universe of creatures to exist since the universe of creatures does in fact exist, it must have this necessity either of itself or from another. But the universe cannot have necessary existence from itself because as demonstrated every being must be from a first being (a first mover, a first efficient cause, or a necessary being). Since that which has being not from itself cannot possibly have necessity of being from itself since that which has being not from itself has the possibility of not being, and what must necessarily be, cannot possibly not be. Consequently, that which of itself has necessary being, has of itself the impossibility of not being and because it has of itself the impossibility of not being, it is not a non-being. Therefore the necessary being since it has of itself the impossibility of not being and is therefore not a non-being, the necessary being must necessarily exist. In modal necessity of this type, the consequent of impossibility is necessity and not possibility. Thus one cannot say the possibility of being implied being, and therefore being implies necessary existence. However, in this modal necessity one can say, impossibility of not-being implies the necessity of being.

things that are possible to be and not to be, since they are found to be generated, and to corrupt, and consequently, they are possible to be and not to be," the argument is likewise *synthetic*. The conditional necessity is abstracted from the nature of things, and thus the necessity itself is *synthetic* while being necessary and universal since the natural necessity abstracted is abstracted from sensible things as universal. Thus one can attain a synthetic a *priori* natural knowledge that the Necessary Being exists and by *naturaliter nota* this Necessary Being is considered to be God.

Chapter 7

CONCLUSION

Universal and necessary *a priori* synthetic propositions are possible and Kant's program was in fact an unnecessary development of early modern philosophy. Kant argued in the Critique: "the proper problem of pure reason is contained in the question: How are a priori synthetic judgments possible? That metaphysics has remained in uncertainty and contradiction is due to this problem and because no one has considered the distinction between analytic and synthetic judgments. Upon the solution to this problem, or upon a sufficient proof that the possibility which it desires to have explained does in fact not exist at all, depends the success or failure of metaphysics."[494]

The thesis concludes using the historical critical method and a proper reformulation of the synthetic *a priori* that the *possibility* of a solution which the Kantian transcendental philosophy desired to explain does in fact exist, but simply not in the manner in which the Kantian solution proposed. The sufficient proof demonstrates that the problem itself that Kant was attempting to address simply does not exist at all. Kant left one with two options, either the transcendental solution or no solution at all. However, this is a false dichotomy since there are other possible solutions to the problem of the *a posteriori*/synthetic *a priori* distinction beyond those which Kant and Hume considered.

---

494 Immanuel Kant, *Critique of Pure Reason*, trans. by N. Kemp Smith, (London: Macmillan, 1929), 55.

The thesis has attempted to develop a cogent reformulated synthetic *a priori* solution concluding that it is a far more appropriate or fitting response to the Kantian synthetic *a priori* than proposing an *a posteriori* solution when it comes to resolving issues of the Kantian transcendental philosophy.[495] The Kantian program was introduced by the debates between the British Empiricists and the Continental Rationalists, which were historically and fundamentally flawed, introducing a false dichotomy between sensibility and intelligibility. The debate introduced a mistaken disparity between the ontological status of causality and an empirical knowledge of causal necessity as well as between sensibility and intelligibility or between the *a posteriori* and the *a priori*.

Kant influenced by the Continental Rationalist tradition and shaped by his early Dogmatism was awakened from his Dogmatic slumber through exposure to Newtonian empiricism and Hume's empiricist critique. However, Kant still held to the Dogmatic method even during his critical period allowing it to influence the form and development of the transcendental philosophy. Hume as a British Empiricist and Kant as influenced by both Dogmatism and Empiricism, necessarily followed the disparity between sensibility and

---

495 Although Saul Kripke has argued that certain modal propositions are a posteriori, it should now be clear that the reason why certain propositions are both a priori and a posteriori is because there is no false dichotomy between sensibility and intelligibility. Therefore the proposition, "If the lectern is not made of ice, then necessarily it is not made of ice," provides a posteriori knowledge to the extent that the a posteriori is not detached from the a priori since the proposition is both necessary from experience and necessary in reason since if something is made of ice, it is necessarily made of ice both in reason and experience. As Albert Casullo observed in his "Reply to My Critics: Anthony Brueckner and Robin Jeshion" during the March 2005 Pacific APA conference: "it seems if we have available to us the two different ways of coming to understand each of the sentences, and each way of understanding each of the sentences opens up a different way of coming to know that the sentences are true, then we have the possibility of knowing each of those sentences both *a priori* and *a posteriori*... So, it appears that if we consider the three Kripke sentences that are in play here—the lectern sentence, the heavenly-body sentence, and the meter-stick sentence—each can be known either a priori or a posteriori depending on how one comes to understand the sentence. It follows, then, that it is not the case that the first two are knowable only a posteriori." I agree with Albert Casullo's observation and his critique, but the reason why such propositions are not knowable only a posteriori is because the a posteriori is not as Hume held independent of reason and likewise the a priori is not independent of experience. Hence the early modern dichotomy was a false problem introduced by early modern Humean psychology and Kantian transcendental philosophy as well as the developments of early modern Continental Dogmatism and British Empiricism.

intelligibility developed by early modern philosophy in their respective solutions. Although Kant's solution was an attempt at a compromise between British Empiricism and Continental Rationalism after Hume, transcendental idealism was essentially a Continental Rationalist development in reaction to the British Empiricism of Hume.

Having developed a cogent argument that universal and necessary knowledge of efficient causality can be attained from experience independent of the transcendental philosophy, it is the conclusion of this book that one can demonstrate God's existence from knowledge of modal necessity and efficient causality laying a philosophical foundation for an integrated model for an intelligible and sensible knowledge of God's existence through a reformulated synthetic *a priori* solution. However, one might ask why not a reformulated *a posteriori* solution rather than a reformulated synthetic *a priori* proposal?

For Aquinas the mind is in potency to intelligible objects and these objects are not actual in the mind until the object is understood by the intellect.[496] The intellect according to Aquinas is like a blank sheet of paper on which no word is yet written, but the intellect is in potency to all things.[497] Moreover, the proper object of the intellect is what a thing is which is its essence. What is known of a thing is known by knowing its substance and this includes knowing the accidents either by discursive reason or by the senses perceiving the accidents of a thing and the cognitive faculty apprehending that what is perceived in/as this or that particular substance, and by this one can arrive at the substance and know the essence of a thing.[498]

---

496 Aquinas indicates: "The mind, then, is called passive just in so far as it is in potency, somehow, to intelligible objects which are not actual in it until understood by it. It is like a sheet of paper on which no word is yet written, but many can be written. Such is the condition of the intellect as a potency, so long as it lacks actual knowledge of intelligible objects. This is against, not only the early natural philosophers' view that the soul knows all things because it is composed of all things, but also Plato's opinion that the human soul is by nature in possession of a universal knowledge which only its union with the body has caused it to forget. (This theory is implicit in Plato's reduction of learning to remembering.)" (*In De Anima*, III, Lec 9 Sct 723).

497 Ibid.

498 Aquinas maintains: "No cognitive power knows a thing except under the aspect of its proper object: thus by sight we do not know a thing except as colored. Now the proper object of the intellect is what a thing is, namely the essence of a thing, as stated in 3 De Anima, iv. Consequently whatever the intellect

Aquinas also understood, following Aristotle, the problem suggested by the distinction between sensibility and intelligibility. For the intellect to be all things, the intellect must be immaterial and unmixed. However, the senses are mixed and material and are limited to those things that sense is capable of sensing.[499] Aquinas also maintained that the truth determined in judgment is essentially conformity or correspondence between a thing and the intellect through assimilation of the knower to the thing known according to the mode of the knower.[500] Thus the question arose, how does this conformity

knows of a thing, it knows it through the knowledge of its essence, so that whenever by demonstration we become acquainted with the proper accidents of a thing, we take as principle, what that thing is, as stated in 1 Poster. i. iv. On the other hand, if the intellect knows the essence from its accidents, according to the statement in 1 De Anima, i. that accidents are a great help in knowing what a thing is; this is accidental, in so far as the knowledge of the intellect arises from the senses, and so by knowing the accidents as perceived by the senses we need to arrive at knowing the substance: for this reason this does not occur in mathematics, but only in physics. Consequently whatever cannot be known in a thing by knowing its substance must be unknown to the intellect" (*SCG* IIIa, 56).

499 Aquinas argues: "But this led to the exclusion of the opinion of Empedocles and other ancient thinkers who stated that the knower is of the nature of the thing known, for example that we know earth inasmuch as we are earth, and water inasmuch as we are water. But Aristotle showed above that this is not true of sense because a sense power is not actually but potentially those things that it senses; and here he says the same thing of the intellect. But there is a difference between sense and intellect, because a sense is not able to know all things, but sight can know only colors; hearing, only sounds; and so for the rest; whereas the intellect is able to know all things without such limitations. Now the ancient philosophers used to say, since they were of the opinion that the knower must have the nature of the thing known, that the soul, in order to know all things, must be a mixture of the principles of all things. But because Aristotle already proved through a comparison with sense, that the intellect is not actually but only potentially that which it knows, he concluded to the contrary that "because the intellect knows all things, it must be unmixed," that is, not composed of all things as Empedocles had stated" (*On the Unity of the Intellect against the Averroists*, I Sct 19).

500 Aquinas explains: "True expresses the correspondence of being to the knowing power, for all knowing is produced by an assimilation of the knower to the thing known, so that assimilation is said to be the cause of knowledge. Similarly, the sense of sight knows a color by being informed with a species of the color. The first reference of being to the intellect, therefore, consists in its agreement with the intellect. This agreement is called "the conformity of thing and intellect." In this conformity is fulfilled the formal constituent of the true, and this is what the true adds to being, namely, the conformity or equation of thing and intellect. As we said, the knowledge of a thing is a consequence of this conformity; therefore, it is an effect of truth, even though the fact that the thing is a being is prior to its truth" (*De Veritate* I, q. 1, a. 1, body). Likewise, "The human intellect must of necessity understand by composition and division. For since the intellect passes from potentiality to act, it has a likeness to things which are generated, which do not attain to perfection all at once but acquire it by degrees: so likewise the human intellect does not acquire perfect knowledge by the first act of apprehension; but it first apprehends something about its object, such as its quiddity, and this is its first and proper object; and then it understands the properties, accidents, and the various relations of the essence. Thus it necessarily compares one thing with another by composition or division; and from one composition and division it proceeds to another, which is the process of reasoning....

....The likeness of a thing is received into the intellect according to the mode of the intellect, not

arise between sensibility and intelligibility? Aquinas's solution, unlike the transcendental philosophy, avoided a false dichotomy between sensibility and intelligibility that arose following Cartesian dualism and nominalism in the early modern period and that continues today among certain contemporary philosophers.

Finally Aquinas maintained that the intellect can know both universal and necessary things as well as those things that are contingent. The intellect can know the contingent either as contingent, or as containing some element of necessity, since every contingent thing has in it something necessary. In knowing the contingent as contingent, the contingent is directly known by sense and indirectly known by the intellect, while the universal and necessary principles of contingent things are known only by the intellect.[501] In Aquinas, knowledge of

according to the mode of the thing. Wherefore something on the part of the thing corresponds to the composition and division of the intellect; but it does not exist in the same way in the intellect and in the thing. For the proper object of the human intellect is the quiddity of a material thing, which comes under the action of the senses and the imagination. Now in a material thing there is a twofold composition. First, there is the composition of form with matter; and to this corresponds that composition of the intellect whereby the universal whole is predicated of its part: for the genus is derived from common matter, while the difference that completes the species is derived from the form, and the particular from individual matter. The second comparison is of accident with subject: and to this real composition corresponds that composition of the intellect, whereby accident is predicated of subject, as when we say "the man is white." Nevertheless composition of the intellect differs from composition of things; for in the latter the things are diverse, whereas composition of the intellect is a sign of the identity of the components. For the above composition of the intellect does not imply that "man" and "whiteness" are identical, but the assertion, "the man is white," means that "the man is something having whiteness": and the subject, which is a man, is identified with a subject having whiteness. It is the same with the composition of form and matter: for animal signifies that which has a sensitive nature; rational, that which has an intellectual nature; man, that which has both; and Socrates that which has all these things together with individual matter; and according to this kind of identity our intellect predicates the composition of one thing with another." (*ST* I, q. 85, a. 5).

501 In answering whether the intellect can know contingent things, Aquinas indicates:

"OBJ 1: It would seem that the intellect cannot know contingent things: because, as the Philosopher says (Ethic. vi, 6), the objects of understanding, wisdom and knowledge are not contingent, but necessary things.

OBJ 2: Further, as stated in Phys. iv, 12, "what sometimes is and sometimes is not, is measured by time." Now the intellect abstracts from time, and from other material conditions. Therefore, as it is proper to a contingent thing sometime to be and sometime not to be, it seems that contingent things are not known by the intellect.

On the contrary, All knowledge is in the intellect. But some sciences are of the contingent things, as the moral sciences, the objects of which are human actions subject to free-will; and again, the natural sciences in as far as they relate to things generated and corruptible. Therefore the intellect knows contingent things.

I answer that, Contingent things can be considered in two ways: either as contingent, or as containing

efficient causality can be apprehended as an incidental sensible by the cogitative power while its universality and necessity is abstracted from sensibility by the intellect.

In answering the original question, rather than arguing that the *a posteriori* is universal and necessary, it has been argued that synthetic judgments can be arrived at from experience following the noetic of Aristotle and Aquinas and such synthetic judgments are both universal and necessary making them *a priori* but independent of the transcendental philosophy. Independent of the transcendental philosophy in the sense of being independent of the rules for the imaginative synthesis of the manifold or simply independent of the rules or schemata some of which are empirical and others are *a priori*. Although Kant argued that the *a priori* is independent of experience, the *a priori* was considered independent of experience due primarily to Hume's critique and it has cogently been argued that this critique itself was a false development in the long history of philosophy.

To address Kant's original disparity as historically determined by Cartesian dualism and the British Empiricist and Continental Rationalist developments in psychology and perception, one must resolve the issue at the level of synthetic judgments in addressing Kant's noetic and Hume's critique against the *a priori* because it is the synthetic *a priori* that Kant associates with both possible experience and universality and necessity. For Kant like Hume the *a posteriori* is limited <u>to empirical intuitions independent of either universality or necessity.</u>

<u>some element of necessity, since every contingent thing has in it something necessary</u>: for example, that Socrates runs, is in itself contingent; but the relation of running to motion is necessary, for it is necessary that Socrates move if he runs. Now contingency arises from matter, for contingency is a potentiality to be or not to be, and potentiality belongs to matter; whereas necessity results from form, because whatever is consequent on form is of necessity in the subject. But matter is the individualizing principle: whereas the universal comes from the abstraction of the form from the particular matter. Moreover it was laid down above (A1) that <u>the intellect of itself and directly has the universal for its object; while the object of sense is the singular, which in a certain way is the indirect object of the intellect,</u> as we have said above (*ST* I, q. 86, a. 1). Therefore <u>the contingent, considered as such, is known directly by sense and indirectly by the intellect; while the universal and necessary principles of contingent things are known only by the intellect.</u> Hence if we consider the objects of science in their universal principles, then all science is of necessary things. But if we consider the things themselves, thus some sciences are of necessary things, some of contingent things. From which the replies to the objections are clear." (*ST* I, q. 86, a. 3).

Analytic judgments for Kant are judgments where the predicate is contained in the subject and are therefore independent of experience because one can deduce the predicate from the subject independent of experience. However, in the case of synthetic judgment, such judgments are extensions of the subject and Kant admits that synthetic judgments are associated with experience in some way unlike analytic judgments that have no need of experience since the predicate is contained in the subject. Kant developed the transcendental philosophy specifically to address how synthetic judgments associated in some way with experience could be *a priori* because Hume had made it impossible for Kant to use the *a posteriori*. So for Kant the *a posteriori* became empirical, associated not with intelligibility but with experience, and thus Kant made the *a posteriori* essentially a meaningless term following Hume when it comes to discursive reason and judgment, and in regard to the *a priori*.

For Kant the *a posteriori* became associated with and a synonym for Hume's notion of custom independent of reason. Hume felt we could infer causality only from experience of similar or like instances of causal events and these similarities established custom from which one could believe and thereby intuitively rather than discursively infer the causal maxim or any notion of causality, and this inference of course was entirely independent of reason or argumentation. Kant realized the implications for science and mathematics. Efficient causality, causality in general, nor substance nor any of the other categories could be apprehended from experience by the intellect since the intellect was clearly detached from experience as demonstrated by Hume. Kant awaking from his Rationalist slumber proposed a Rationalist account in opposition to Hume's Empiricist approach.

Kant being essentially a Continental Rationalist and having accepted Hume's logical conclusion of the disparity between intelligibility and experience that had developed between the British Empiricists and Continental Rationalists, developed the synthetic as a basis to explain how causality, substance, and other notions could be associated with experience but not derived at from experience. It was of course an ingenious move on the part of Kant since it circumvented the issue of the *a posteriori* all together, but it furthered the already existing false dichotomy between sensibility and intelligibility that had developed between the Continental Rationalists and the British Empiricists. The *a posteriori* for neither Hume nor Kant could give universality or necessity, and thus the *a posteriori* became a meaningless term in this regard. The fundamental issue for Kant was how universal and necessary synthetic judgments—particularly of efficient causality—could be associated with experience without being derived directly from it. Hume's empiricist critique, and the broader developments that culminated in it, had made such derivation impossible.

Those, like Kripke, who rely on the a *posteriori* for the notion of rigid designators, fail to recognize the implications of the Kantian–Humean dichotomy. **Hume's critique** argued experience gives us only "custom" or "habit," never necessity. Thus, for Hume, the a posteriori cannot yield universality or necessity. **Kant's** to preserve science and mathematics, he argued that universality and necessity come from *a priori* synthetic judgments, not from experience. For both thinkers, then, the a *posteriori* is essentially stripped of any claim to necessity. Kripke, in *Naming and Necessity* (1970s), introduced **rigid designators:** terms that refer to the same object in all possible worlds (e.g., "Hesperus = Phosphorus," or "water = H_2O"). His insight was that some statements discovered *empirically* (a posteriori) nonetheless express **necessary truths** once reference is fixed by rigid designation. In other words, Kripke reintroduced **necessity into the *a posteriori*** in a way that neither Hume nor Kant would have allowed.

Kripke's project, then, **sidesteps the problem** rather than resolving it: he presupposes that empirical discoveries (e.g., "water is H_2O") can reveal metaphysically necessary truths, without engaging the deeper question of how experience could ever ground necessity in the first place. He never addressed the fundamental issue of Being and

Necessity. In this sense, Kripke may be said to "fail to recognize the implications" of the Kantian–Humean dichotomy: he bypasses the very critique that led Kant to distinguish between analytic/synthetic and a priori/a posteriori in the first place.

Kripke's necessary a *posteriori* is brilliant but vulnerable: if the Kant–Hume dichotomy holds, then no *a posteriori* claim, however rigidly designated, can bear necessity. Aquinas or Aristotle (and this entire text) would counter: necessity can arise from experience, but only when the intellect abstracts the universal and necessary from what is given in sense. That bypasses both Kant's transcendentalism and Hume's skepticism, while also challenging Kripke's uncritical revival of necessity in the *a posteriori*. Kripke fails to grapple with the Kant–Hume legacy. He assumes necessity can attach to empirical discoveries without explaining how this avoids Hume's critique or Kant's insistence that necessity belongs only to the *a priori*.

Due to the continued disparity between sensibility and intelligibility that was never adequately addressed by Kant and Hume's account, the issue of this thesis became: how can one have synthetic judgments of efficient causality that are universal and necessary, but apprehended from experience in opposition to Kant's transcendental notion of the synthetic *a priori*. The solution was to address directly Kant and Hume using the notion of the synthetic *a priori* since the synthetic is associated in some way with experience or the empirical while the *a priori* notion is associated in some way with universality and necessity where Kant attempts to demonstrate that one can have synthetic judgments of possible experience through the *a priori*. The proposed solution addresses Kant's original problem as developed by Hume in terms of Kant's own solution.

The *a posteriori* is meaningless for Kant because one cannot have synthetic universal and necessary principles or knowledge of efficient causality or substance from the *a posteriori* and therefore the *a posteriori* can neither save science nor metaphysics. Therefore it became obvious that the only way to unravel the false dichotomy or disparity between sensibility and intelligibility or experience and reason was to demonstrate that one can have universal and necessary synthetic judgments or knowledge of efficient causality and substance from experience, but such knowledge is not *a posteriori* because the *a posteriori* is a meaningless term in relation to reason or the *a priori* for Hume and Kant.

For Hume the *a posteriori* is essentially synonymous with custom and experience rather than the intellect or demonstration or discursive reason or judgment. Kant agrees and thus proposes the synthetic/analytic and the *a priori/a posteriori* distinctions in an attempt to circumvent the issue while proposing an awakened Dogmatist solution. Kant was essentially a Rationalist and his solution was that of a Continental Rationalist influenced by Humean British Empiricism. Kant's conception of *a priori* knowledge provided a conceptual connection between necessity, efficient causality, and possible experience. Hence Kant proposed a rationalist solution to solve the dichotomy using the transcendental philosophy of the *a priori*, but Kant simply had no notion of contingent *a priori* knowledge.[502]

Admittedly, Aquinas had no notion of Kant or Hume's *a posteriori/a priori* distinction since Aquinas following Aristotle maintained that one can abstract universal and necessary simple quiddities from experience and the intellect can be assimilated to things. Neither in Kant nor Hume can the *a priori* be abstract from experience using the notion of the *a posteriori*. In Kant it is only analytic and synthetic *a priori* judgments that are universal and necessary and synthetic judgments are associated with possible experience through the transcendental. This relationship between the universal and the necessary with that of possible experience was not allowed in the Humean or Kantian *a posteriori* account. In Hume's account the best one can arrive at *a posteriori* is a feeling, belief, or intuition that follows from custom.

---

502 Possibly in opposition to this view see Albert Casullo, *A Priori Justification*, (Oxford: Oxford University Press, 2003), 202-209.

Similarly, for Kant, one arrives at empirical intuitions from possible experience *a posteriori*.

For Aquinas, like Hume and Kant, sense faculties apprehend experience. However, the sensible *a posteriori*, as understood by Hume, Kant, and in a sense by Aquinas himself, never apprehends universality and necessity. Universality and necessity are apprehended by the intellect alone for Aquinas as abstracted from experience. For Hume any such notion is simply similarity that produces custom, custom generates belief, and from belief one infers such notions as universality and necessity independent of reason or argument. For Kant, the *a posteriori* could provide no notion of universality or necessity following Hume's critique. Therefore it was necessary for this thesis to address the transcendental philosophy or Kant's noetic directly by addressing the original issue facing Kant rather than attempting to force Aquinas to become that which neither he nor Hume nor Kant would have accepted.

Although Kant, like Aquinas and Aristotle, held that experience in and of itself can never directly attain universals and necessity independent of intelligibility as Hume had maintained in his psychology of belief, and thus the *a posteriori* for both Kant and Aquinas is limited to sensibility or experience if the *a posteriori* is taken in the Humean sense. However, in point of fact Aquinas had no such Humean *a posteriori* notion. For Aquinas there was no dichotomy between experience and intelligibility due to a historical philosophical critique of Avicenna and Averroës. For Aquinas what was apprehended in singularity at the level of experience was abstracted in its universality at the level of intelligibility eliminating entirely a limited Humean notion of the *a posteriori*.[503]

---

503 Casullo correctly notes the divide in contemporary epistemology between those who embrace and those who reject the *a priori*. The solution to the divide is the solution embraced by the *via antiqua* where there is no *a priori/a posteriori* incongruity. There is formal and material logic, but there is no dichotomy or disparity between the two. Casullo describes the issue at hand: "the major divide in contemporary epistemology is between those who embrace and those who reject the a priori. The importance of the issue, however, extends beyond the boundaries of epistemology to virtually every other area of philosophy. To a large extent, one's views about the a priori determine how one goes about answering other philosophical questions. Current opinion is deeply divided and radically polarized. Proponents of the a priori frequently allege that rejecting it is tantamount to rejecting philosophy as

The *historical critical approach* addressing the original problematic that Kant was attempting to address because of Hume's critique has cogently demonstrated that Kant's problem indeed was mistaken following an unnecessary false dichotomy between experience and reason introduced by early modern philosophy. The solution provided by the thesis avoids the unnecessary false dichotomy and disparity between sensibility and intelligibility, addressing Kant's mistaken solution to a false problem where Kant introduced the synthetic *a priori as an attempt to solve a false disparity between intelligibility and sensibility*. If the thesis were to critique specifically Hume, the author might have appropriately selected the *a posteriori* since there is no clear notion of the *a priori* synthetic/*a posteriori* distinction in Hume, but the author addresses Kant and in reference to Kant the problem lies with the synthetic *a priori* because the synthetic *a priori* is the Kantian solution to the false dichotomy between sensibility and intelligibility introduced by early modern philosophical developments.

If one were to address only Hume, one would be forced, like Kant, to propose that beyond Hume's notion of the *a posteriori* which is limited only to experience and independent of reason, that there is also intelligibility whereby universality and necessity are attainable, and in opposition to Kant it would be necessary to add that this universality and necessity can be abstracted from experience. This is essentially the Kantian notion of the synthetic *a priori* but independent of the transcendental. Independent of the transcendental philosophy in the sense of being independent of the rules for the imaginative synthesis of the manifold or simply independent of the rules or schemata some of which are empirical and others are *a priori*.

---

a respectable intellectual discipline. Opponents respond that no intellectually respectable theory of knowledge can accommodate the a priori" (Albert Casullo, *A Priori Justification*, (Oxford: Oxford University Press, 2003), 3). Rather than attempting to justify the Kantian *a priori*, the proposed critique of pure reason has been an attempt at finding a middle ground between deeply divided opinions that are often radically polarized. A middle ground has been suggested by cogently demonstrating that the disparity between intelligibility and sensibility was in fact an unnecessary philosophical development of the early modern period due to a loss of ancient and medieval developments in cognitive psychology and philosophical anthropology.

However, since the thesis is addressing Kant rather than Hume, it was necessary to address Kant's problematic and Kant's solution which is the synthetic *a priori* rather than the *a posteriori*. Therefore it was necessary to propose the synthetic *a priori* independent of the transcendental as the proper solution to the early modern dichotomy between sensibility and intelligibility existing contrary to the natural mode of human knowability. In any case, one seems to be led to the proper conclusion that because of the false dichotomy between sensibility and intelligibility and the association of the *a posteriori* with experience independent of intelligibility in early modern philosophy, the synthetic *a priori* independent of the transcendental appropriately represents the proper order of knowability from sensibility to intelligibility. Although it has been alleged that our thesis circumvents the central issue of Hume's problematic. But it does in fact address directly Hume's central thesis and Kant's response to Hume in developing a critique of Kant's notion of the synthetic *a priori* while cogently arguing that the transcendental was itself simply an unnecessary early modern innovation. Hume was simply mistaken as was Immanuel Kant. Kant should never have awakened from his dogmatic slumber. The Kantian Copernican Revolution was unnecessary as was the analytic/synthetic divide that followed.

The book has made a cogent and what seems to be a sound case, if not demonstrative, that one can know following the *via antiqua* noetic in opposition to that of the *via moderna* that universality and necessity is possible as abstracted from experience contrary to the Kantian transcendental philosophy and the Humean critique. If this is the case, it seems to follow from the our thesis that it is possible to attain synthetic *a priori* knowledge of efficient causality. It follows, once one apprehends synthetic *a priori* knowledge of efficient causality and necessity from experience, one can likewise argue for the necessary existence of God from natural or modal necessity and efficient causality following the arguments from efficient causality and necessity. A priori arguments do not have to analytic nor detached from experience, and this is likewise true of symbolic modal logic such as Krypke's possible worlds and frege's symbolic logic. It follows from such demonstrations that God necessarily exists. Therefore, it has been reasonably demonstrated using synthetic *a priori* efficient causality that God necessarily exists. The second book in this series will lay the foundation of numerous synthetic *a priori* formal arguments to this effect.

GLOSSARY

This list of definitions is intended for readers unfamiliar with Scholastic terminology. In many cases, the definitions have been simplified and in the majority of cases are adopted from the *The Disputed Questions on Truth*, vol. 2 Glossary trans. by James V. McGlynn, S.J. (Chicago: Henry Regnery Co., 1953) and W. Norris Clarke, S.J., *The One and the Many: A Contemporary Thomistic Metaphysic, Glossary*, (Notre Dame: University of Notre Dame Press, 2001). The Glossary does not attempt to define various technical terms developed by Kant as understood by various commentators.

Abstract, Abstraction (n.). The content of a given idea or concept signifying some aspect of their object but leaving out others. Universal ideas that abstract the essence or the defining notes rather than simply the subjective meaning of something while omitting the concrete individual details. For example, abstracting from the particular details of being Socrates the more general concept of man.

Absolute (n.). A being that stands or is conceived in itself and not in reference to something else; opposed to what is relative, e.g., man is an absolute whereas father is a relative.

Absolutely (adv.). Without regard to any particular circumstance; separated from all that is not itself, e.g., man considered absolutely is man considered simply and solely as rational animal. Similarly, that which has necessity from the efficient cause is absolutely necessary. Thus because of the motion or rotation of the earth it is necessary that day and night alternate. *In Physics* II, 15, 270

Accident (n.). That which inheres in and in relation to a substance, i.e., exists in it as a modification, e.g., color with reference to the thing colored; place with reference to the thing located.

Act (n.). A perfection, as that of existence; an actuality.

Analogy, Analogous (n.). The property of a concept by which a term is predicated of several different subjects according to meaning partly the same, partly different. 1. Analogy of extrinsic attribution: an analogous term is predicated properly only of one primary analogate and of the secondary analogates only due to some relation such as causality, but not because of some intrinsic similarity (e.g., healthy food and healthy woman). 2. Analogy of proportionality where the analogous term is predicated because of some proportional similarity between the analogate: a) Improper proportionality where the analogous term is properly predicated of one analogate and of the other metaphorically (e.g., an angry man and the angry sea), b) proper proportionality where the analogous term is predicated properly of all the analogates (e.g., a man knows and God knows).

Appetite (n.). 1. General: The faculties of desire. 2. Rational appetite: The will. 3. Sensitive appetite: The faculties by which one is inclined to seek what is suitable for the senses and to flee from what is harmful to them (the concupiscible power), or to resist whatever opposes the objects of the concupiscible (the irascible power) cf. *ST* I, q. 81, a. 2. 4. Natural appetite: The natural inclination flowing from the form or nature which naturally determines the proper being of a thing (cf. *ST* I, q. 80, a. 1).

Art (n.). A form or plan in the intellect of an artist, according to which he makes something.

Being (n.). That whose act is to be; that which is defined by a reference of whatever sort to existence. Briefly, any subject of existence. 1. Real being that which has its own act of existence outside of an idea or knowing subject. 2. Intentional being that which has intentional being in reason or understanding insofar as its being is its to-be-thought-about for example numbers, abstract thought, hypotheses, etc.

Cause (n.). 1. (General): That which gives existence to another. 2. Appropriated cause: Same as proper cause. See (12), below. 3. Common cause: See (12), below. 4. Efficient cause: The extrinsic principle which gives existence. The source of motion as by an agent. "For if every process of generation and corruption is from some one thing or more than one, why does this occur, and what is the cause? For certainly the subject itself does not cause itself to change. I mean, for example, that neither wood nor bronze is the cause of the change undergone by either one of them; for wood does not produce a bed, or bronze a statue, but something else is the cause of the change. But to seek this is to seek another principle, as if one were to say that from which the beginning of motion comes" (*Aquinas: Meta.* I, Bk. 1, Lsn. 5, Sct. 45). "In a third sense cause means that from which the first beginning of change or of rest comes, i.e., a moving or efficient cause. He says "of change or of rest," because motion and rest which are natural are traced back to the same cause, and the same is true of motion and of rest which are a result of force. For that cause by which something is moved to a place is the same as that by which it is made to rest there. "An adviser" is an example of this kind of cause, for it is as a result of an adviser that motion begins in the one who acts upon his advice for the sake of safeguarding something. And in a similar way "a father is the cause of a child." In these two examples Aristotle touches upon the two principles of motion from which all things come to be, namely, purpose in the case of an adviser, and nature in the case of a father. And in general every maker is a cause of the thing made and every changer a cause of the thing changed." (*Aquinas: In Meta.* I, Bk. 5, Lsn. 2, Sct. 765). 5. Equivocal cause: An efficient cause whose effect is specifically different from itself; as distinguished from a univocal cause, whose effect is specifically the same as itself. 6. Exemplary cause: A form conceived in the mind of a free agent that serves as a model for the production of a given effect. 7. Final cause: That on account of which something is or is done; the end or purpose; the thing which incites, moves, and determines the efficient cause by attraction; some good which motivates the agent's activity. 8. Formal cause: The constituent principle that accounts for the specific perfection of a composite being, e.g., the soul of man. 9. Instrumental cause: A type of efficient cause that exercises its causal function under the directive influence of an agent or principal cause, thereby producing an effect that exceeds its unaided powers of production, e.g., a pen in the hand of a poet. 10. Material cause: The constitutive potential principle of a composite being, e.g., the marble of a statue. 11. Particular cause: An efficient cause whose productive activity is restricted to this or that particular class of effects; as distinguished from a universal cause, i.e., an efficient cause whose productive activity is not thus restricted. 12. Proper cause: In creatures, a cause which is determined to one effect and one only; as distinguished from common cause, i.e., a cause whose causality is not determined to one effect. 13. Proximate cause: A cause that produces its effect directly without any other cause intervening; as distinguished from remote cause, i.e., a cause which produces its effect mediately, through other intervening causes. 14. Remote cause: See (13), above. 15. Universal cause: See (11), above. 16. Univocal cause.: See (5), above.

Character, Intelligible (n.). A nature, essence, or note as knowable.

Commentator, The (n.). Averroës (1126-1198). Arabian commentator on Aristotle's works.

Composed (adj.). Made of parts or explicitly conceived as having parts; as distinguished from non-composed, i.e., not made up of parts or not conceived as having distinct parts.

Composite (adj.). 1. Made up of parts; compounded. 2. Joined in thought, as in a judgment.

Composition (n.). 1. The act of joining, as in a judgment. 2. The state of being joined. 3. Something joined.

Contraries (n.). Things most opposed to each other in some genus, e.g., immaterial and material.

Difference (n.). (In some contexts) Same as specific difference, that determination added to the generic nature which distinguishes a given species from all other species of the same genus.

Disposition (n.). 1. A modification of a substance, easily changed. 2. The state of a substance ready to receive a new form.

Divide (v.). To deny a predicate of a subject in a judgment.

Element (n.). A primary physical ingredient of things. (The elements were thought to be fire, air, water, and earth.)

Essence (n.). That by which something is what it is; that which is designated by the definition; that which is defined by reference to the primary act of existence; *what a thing is* in contrast to *that which is.*

Estimation (n.). 1. A general evaluation. 2. A judgment. As in the estimative power in animals or the cogitative power unique to humans.

Exemplar (n.). A form in imitation of which a thing comes into being from the intention of a free agent.

Existence (n.). The actuation of the essence; that by which something is or exists; the fundamental act of any being as such; *that which is* in contrast to *what a thing is.*

Faith (n.). 1. A supernatural assent of the intellect, at the command of the will and under the influence of grace, to a revealed truth because of the authority of God who reveals it.

Fallacy (n.). 1. F. of Accident: An argument based on reasoning from what is accidental to a thing as though it were essential to it; the acceptance of mere material identity for formal identity. 2. F. of the Consequent: An illegitimate argument, found usually in a conditional syllogism. It happens in two ways: either by arguing from the falsity of a condition to the falsity of the conditioned clause or from the truth of the conditioned clause to the truth of the condition, as in the following: If it is raining, the ground is moist. But the ground is moist. Therefore, it is raining.

Form (n.). 1. Physical f.: Same as formal cause. See under cause. 2. Intelligible f.: An immaterial representation of the thing known in the intellect of the knower. 3. Separated f.: A separated substance. See under substance.

Form of empirical intuition: 1. formal features or structure of the objects intuited equivalent to 'form of appearances.' 2. the manner or mode of intuiting.

Form of appearances: a feature of the appearance or empirical intuition in virtue of which their elements are viewed as ordered or related to one another in possible experience.

Formally (adv.). According to the proper essential definition of a given thing, e.g., formally, man is a rational animal.

Grace (n.). 1. (General): A supernatural gift of God to a rational creature for the purpose of eternal salvation. 2. Actual g.: A supernatural transient aid conferred by God to elicit supernatural acts. 3. Sanctifying g.: A supernatural permanent gift inherent in the soul, giving it a share in the divine nature without identifying it with that nature.

Habit (n.). 1. General: A modification of a substance, not easily changed; a quality whereby a thing is disposed, either in itself or in relation to something else; an abiding disposition. 2. Infused h.: A habit given with a nature or gratuitously by God. See S.T., I-II, 51, 1 and 4. 3. Acquired h.: A habit which is the result of repeated acts.

Hylemorphism (n.). The composition of all material beings by matter (hylé) and form (morphé).

Imperium (n.). An interior act of reason, forbidding or commanding the will.

Informed (adj.). Specified by an intrinsic formal element.

Intellect (n.). 1. (General): The immaterial faculty of knowing, possessed by the soul. 2. Active intellect: A special power of the soul which works on the phantasm, elevates it, and, by its instrumentality, produces in the possible intellect the intelligible species by which the possible intellect is informed and actuated. 3. Agent intellect (*intellectus agens*): Same as active intellect. 4. Possible intellect (*intellectus possibilis*): The power of the soul to receive intelligible forms and to be brought into the act of understanding. 5. Potential intellect: Same as the possible intellect. 6. Passive intellect: Same as particular reason or the cogitative sense in man, which, because of its conjunction with intellect, is, in a way, able to compare and infer and because of its conjunction with sensibility is passively receives sensible forms.

Intentional (adj.). Pertaining to knowledge or representation under the aspect of its "otherness," i.e., as portraying something else; being in one thing but referring to another, e.g., a cognitive form is said to be intentional because, though it is in the knower, it is the form of the thing known and is that by which the intellect and senses are assimilated to the external object itself as an intentional object of the knower.

Intuition (n.). Immediate or direct knowledge of a present object as it is.

Join (v.). To unite; to affirm a predicate of a subject in a judgment.

Judgment, natural (n.). The estimative power. See under power. Also the act of this power.

Knowledge (n.). 1. (General): An immaterial union of knower and known.

Materially (adv.). Basically or fundamentally only; not formally; as that from which something can be formed (e.g., a nature in singular things is materially a universal inasmuch as from it, when conceived, a concept formally universal can be formed).

Matter (n.). 1. (General): An intrinsic capacity for perfection; pure potency. See cause, material. 2. First matter: The first intrinsic and potential principle of a corporeal essence; an intrinsic constituent principle of a body; as distinguished from second matter, i.e., matter already actuated by a substantial form but still with a capacity for a further or different form. 3. Designated matter: Matter actuated and existing with its quantity under its actual dimensions or in potency to a certain quantity and capable of a particular extension; as distinguished from non-designated matter, i.e., matter actuated by form but considered apart from quantity or extension. 4. Second mattere: See (2), above.

Metaphysics (n.). 1. Properly being *qua* being or being in all its generality or all beings insofar as they exist. 2. Secondarily, all things that fall under being in discovering the primitive properties, principles, and causes that govern all being both real and intelligible being insofar as they exist.

Motion (n.). Any change, whether local, quantitative, or qualitative; or, in a wider sense, any reception of a perfection.

Nature (n.). The essence or essential attributes or properties from which a given subject acts. The action flows from their being the sorts of things they are. They are what they are and they can be no other. And their action follows from what they are. Aquinas writes: "the natural philosopher does not seek to know the nature of a stone and of a horse, save for the purpose of knowing the essential properties of those things which he perceives with his senses. ... in like manner the natural philosopher cannot judge perfectly of natural things, unless he knows sensible things." *ST* I, q. 84, a. 8 body

Negation (n.). The absence or denial of a designated perfection.

Participation (n.). An order of relationship between beings insofar as they share in a given property or perfection common to each and as received from a common source. In participating in *esse*, all finite beings participate in their own particular act of existence from a common source that gives *esse* or the act of existence to all.

Passion (n.). Any undergoing or being acted upon; the reception of a perfection.

Patient (n.). The subject of a passion; that which undergoes something or is acted upon. See passion.

Per accidens (adv. phr.). (1) contingently and incidental; apart from and incidental to an essence; by reason of something else. (2) incidental intention as distinguished from per se, i.e., essentially, directly, intrinsically connected with an action, intention, or essence; by reason of what it is in itself. That which is apprehended as an incidental by the cogitative power.

Per se (adv. phr.) See per accidens.

Perfection (n.). 1. A state of completion, relative or absolute, or a state in which nothing is lacking. 2. Something contributing to this completion or well-being; any good possessed or that may be possessed.

Phantasm (n.). An internal sensible representation of a material thing.

Philosopher, The (n.). Aristotle (384-322 B.C.).

Possible (n.). That which can be; whatever has truth or a relation to being; anything whose notion is not intrinsically contradictory.

Potency (n.). 1. (General): Capacity for perfection. 2. Active potency: A capacity for doing; hence, a principle of action. As distinguished from passive potency, i.e., a positive reality between absolute non-being and being in act; a principle or capacity of being acted upon. 3. Natural potency: The capacity rooted in the nature of a thing for perfections proportionate to its substantial nature; as distinguished from obediential potency, i.e., the capacity a creature possesses to be elevated by God to acts or perfections that exceed the proportion of its substantial nature. 4. Obediential potency: See (3), above. 5. Passive potency: See (2), above.

Power (n.). 1. (General): A capacity for, or principle of, action. 2. Cogitative power: See (3), below. 3. Estimative power: A sense power of certain instinctive concrete associations and adaptations and of the perception of concrete relations, such as the suitability of the thing sensed to the sensing animal. In man, a similar but less determined power, operating under the influence of reason, is called the cogitative power or particular reason. 4. Irascible power: One of the sensitive appetites. See under appetite. 5. Concupiscible power: One of the sensitive appetites. See under appetite.

Predicate (n.). 1. (General): That which is affirmed or denied of a subject in a judgment. 2. Essential predicate: A term signifying an attribute or operation that is common between subjects but essential to a given subject. 3. Personal predicate: A term signifying an exclusive property of a given subject.

Principle (n.). 1. (General): Something from which something else either is, becomes, or is known. 2. Source. A source from which other things flow and their intelligibility depends. Can be (a) primitive principle that governs the order of knowing such as the principle of non-contradiction; (b) or that which governs the order of being such as the principle of non-contradiction or the principle of causality where causality is a metaphysical law of a relation that binds together in a necessary connection of intelligibility in the order of being such as the causal maxim *every being that begins to exist requires an efficient cause*; (c) some element on which a metaphysical structure depends such as essence and the act of existence, form and mater in substantial change.

Privation (n.). The absence of a perfection that should be present in a given subject; e.g., blindness is a privation with respect to man.

Property (n.). That which is necessarily consequent upon the essence of a given thing, e.g., mortality with reference to any living organism. An attribute or characteristic that follows immediately from the essence and is necessarily connected with the essence but does not signify the essence itself.

Quality (n.). An accidental form or perfection by which a being is said to be such and such, e.g., bitter, sweet, knowing; an accidental perfection whose ultimate substantial principle is the form. See Aristotle, *Categoriae*, VIII (8b 25 seq.).

Quantity (n.). 1. (General): That accidental form or perfection properly belonging to body as such, whose effect is extension. 2. Quantity of dimensions: Quantity, together with particular dimensions. 3. Virtual quantity.: The extent of a power taken with reference to multiple objects or of a principle with respect to its object.

Quiddity (n.). 1. (General): Same as essence, i.e., that which is expressed by its definition, what the thing is. That which makes it to be what it is, i.e., this particular being distinct from another. 2. Real quiddity is the essential characteristics of a given substance as individuated in this or that subject. It is that which makes this or that subject what it is. 3. Intelligible quiddity is the intelligible or intentional definition or defining notes of what a thing is where the simple quiddity is apprehended in simple apprehension and where the complex quiddity is apprehended in judgment as the predicate is combined or separated with a given subject. For example, the essence of man is a rational creature that walks on two legs and laughs, etc. Further, reason and analysis discursively define the essence or quiddity of a given subject.

Reason (n.). 1. (General): The intellectual power of man, especially as it knows by concluding from premises. 2. Particular reason: The cogitative sense in man, which, because of its conjunction with intellect, is, in a way, able to compare and infer. See power, estimative.

Relation (n.). 1. (General): An order, reference, or proportion of one thing to another. 2. Conceptual or logical relation: A relation which can exist only as an object of thought within the mind that conceives it. E.g., the relation of abstract man to real man. Logical relation is distinguished from real relation, i.e., a relation that exists independently of the mind, such as the relation of an actually existing father to his son. For types of real relations, see (3), (6), below. 3. Predicamental relation: That type of accident, the total nature of which consists in the reference of one thing to another. It is distinguished from transcendental relation, i.e., an essential reference which a principle of being, either actual or potential, has to its correlative. 4. Rational relation: Same as conceptual relation. 5. Real relation: See (2), above. 6. Transcendental relation: See (3), above.

Science (n.). (General) 1. Any certain intellectual knowledge. 2. Certain knowledge drawn from first principles by reasoning, i.e., knowledge through causes.

Sensation (n.). 1. (Act): The act of cognition which takes place when a sensible form is received into the corporeal organ of a sense power. 2. (Power): A power residing in a physical organ, capable of receiving sensible forms without their matter (without, however, changing or destroying their nature), by which forms the act of sense knowledge is had.

Sense (n.). 1. Common sense as distinguished from common sensibles: An internal power of awareness of sensation and of distinguishing between the sensations and objects of the several external senses. 2. Proper sense: An external sense with a special object, e.g., vision, which senses only color.

Sensible (adj.). Capable of being known by a sense power, e.g., color, sound, etc.

Sensible (n.). 1. (General): That which is capable of being known by a sense power. 2. Common sensible: See (4), below. 3. Per se sensible: that which is the object of sensation; as distinguished from a per accidens sensible, which is really an object of intellection but is known by the senses incidentally. For example, what the eye knows as white happens to be Socrates' son; hence, Socrates' son is said to be a sensible per accidens. 4. Proper sensible: That which is the peculiar or special object of a single external sense, e.g., for vision, color, for hearing, sound; as distinguished from a common sensible, which is attained by several senses, as shape or size, being in each case a quantitative aspect of the material thing. The Aristotelian view held that common sensibles, similar to incidental or per accidens sensibles, were less reliably known than proper sensibles due to proper sensibles being a peculiar or special object of a single external sense. The modern turn considers common sensibles to be more reliably known than proper sensibles due to the quantitative nature of common sensibles and are considered "primary qualities." The modern view thus distinguishes between more reliably known "primary qualities" that are common sensibles and less reliably known proper sensibles considered "secondary qualities." After Berkeley critique of primary qualities, both primary and secondary qualities were no longer considered reliable measures of existence and were considered notions or ideas or intuitions about existence.

Signate (adj.). When used of matter, same as designated. See matter.

Simple (adj.). Having no parts; not composed of matter and form, hence, not extended.

Simply (adv.). In the concrete, with all relations and attendant circumstances; without further qualification.

Species (n.). 1. (General): A particular type of being. 2. Intentional species: The cognitive form by which the knowing power is informed and made like something else. 3. Logical species: A common nature considered as apprehended with its distinguishing determination and explicitly referred to many individuals in which it is to be found.

Subject (n.). 1. That which receives a perfection, e.g., substance as regards an accident. 2. (Logical): That of which something is predicated.

Substance (n.). 1. (General): That being, the essence of which is defined by a natural exigency for the primary act of existence, which act it thereby possesses as the ultimate and independent intrinsic subject of being; a being of such a kind as to have existence in and by virtue of itself as an independent intrinsic subject of being. 2. Separated substance: A created intellectual subsistent being, whose essence does not include matter, e.g., an angel. 3. Intelligible substance: Same as separated substance.

Supposite (hypostasis) (n.). An individual, complete substance, existing in itself and not as a part of another.

Understanding (n.). 1. (General) (A.): The act or faculty by which strictly immaterial knowledge takes place. (B.) Intellectual knowledge had without discursive reasoning. (C.): Habitual knowledge of first principles. 2. Knowledge of simple understanding: See knowledge.

Virtually (adv.). Contained in a cause which has the power of producing it, e.g., the warmth of other things is contained virtually in a fire.

AN EFFECT IS CONTAINED VIRTUALLY IN A CAUSE WHICH HAS THE POWER OF PRODUCING IT, E.G., WARMTH IS CONTAINED VIRTUALLY IN FIRE

BIBLIOGRAPHY

Primary Texts

Aquinas, St. Thomas. *Aristotle's De Anima with the Commentary of St. Thomas Aquinas*. Translation by K. Foster and S. Humphries. New Haven: Yale University Press, 1951.

Aristotle on Interpretation: Commentary by St. Thomas and Cajetan. Translation by J. T. Oesterle. Milwaukee: Marquette University Press, 1962.

Commentary on the Metaphysics of Aristotle. Translated by J. P. Rowan. Chicago: Regnery, 1964.

Commentary on Aristotle's Physics. Translation by R. J. Blackwell et. al. New Haven: Yale, 1963.

Commentary on the Posterior Analytics of Aristotle. Translated by F. R. Larcher Albany, NY: Magi Books, 1970.

Compendium of Theology. Translated by Cyril Vollert. Herder: St. Louis, 1947.

The Disputed Questions on Truth. Vol. 1 trans. by Robert William Mulligan S. J.; Vol. 2 trans. by James V. McGlynn, S.J.; Vol. 3 trans. by Robert W. Schmidt, S.J. Chicago: Henry Regnery Co., 1952.

The Disputed Questions on Truth. Chicago: Henry Regnery Co., 1954.

The Division and Methods of Sciences: Questions V and VI of his Commentary on the De Trinitate of Boethius translated with Introduction and Notes, trans. Armand Mauer. Belgium: Universa Wetteren, 1963.

Exposition on the Posterior Analytics of Aristotle. Translated by Pierre H. Conway. Quebec: *La Libraire Philosophique* M. Doyon, 1956.

Faith, Reason, and Theology, Questions I-IV of the Commentary on Boethius' De Trinitate. Translation by Armand Maurer. Toronto: Pontifical Institute of Mediaeval Studies, 1986.

On Being and Essence, translation and introductory notes by Armand A. Maurer, C.S.B. Toronto: Pontifical Institute of Mediaeval Studies, 2nd ed., 1968.

On the Eternity of the World. Translation by Cyril Vollert. Milwaukee: Marquette University Press, 1964.

On the Power of God. Translated by English Dominican Fathers. London: Burns, Oates, and Washbourne, 1932-34.

On the Unity of The Intellect Against the Averroists. Translation and intro. by Beatrice H. Zedler. Milwaukee: Marquette University Press, 1968.

Opera Omnia, iussu Leonis XIII PM edita. Vol. XLIII, De Principiis Naturae, De Ente Et Essentia, and *De Propositionibus Modalibus* by Santa Sabina (Aventino). Roma: Editori di San Tommaso, 1882.

Opera Omnia, iussu Leonis XIII PM edita. Vol. IV, Summa Theologica, Pars Prima by Santa Sabina (Aventino). Roma: Editori di San Tommaso, 1882.

Questions on the Soul, trans. by James H. Robb. Milwaukee: Marquette University Press, 1984.

Scriptum super libros Sententiarum Magistri Petri Lombardi, III, IV, VII ed. R. P. Mandonnet and R. P. M.F. Moos, 4 vols. Paris: Sumptibus P. Lethielleux, Editoris, 1929-47.

Summa Theologiae. Literally translated by English Dominicans. London: Burns, Oates, and Washbourne 1912-36; repr. New York: Benziger 3 vols. 1947-48; repr. New York: Christian Classics, 1981.

Thomas Aquinas A Commentary on Aristotle's De Anima. Translation by Robert Pasnau. New Haven: Yale University, 1999.

Aristotle. *The Complete Works of Aristotle The Revised Oxford Translation* Vol II, *Metaphysics*. Edited by Jonathan Barnes and revised by J. A. Smith. New Jersey: Princeton University Press, 1984. Translated by Ingram Bywater. Oxford: The Clarendon Press, 1928.

The Complete Works of Aristotle The Revised Oxford Translation Vol III, *Metaphysics*. Edited by Jonathan Barnes and revised by J. A. Smith. New Jersey: Princeton University Press, 1984. Translation by W. D. Ross. Oxford: Clarendon Press, 1924.

The Complete Works of Aristotle The Revised Oxford Translation Vol II, *Nicomachean Ethics*. Edited by Jonathan Barne and revised by J. O. Urmson. New Jersey: Princeton University Press, 1984. Translation by Bywater. Oxford: OCT, 1894

The Complete Works of Aristotle The Revised Oxford Translation Vol I, *On the Soul*. Edited. by Jonathan Barnes and revised by J. A. Smith. New Jersey: Princeton University Press, 1984. Translation by W. D. Ross. Oxford: OCT, 1956.

The Complete Works of Aristotle The Revised Oxford Translation Vol I, *Physics*. Edited. by Jonathan Barnes and revised by R. P. Hardie and R. K. Gaye. New Jersey: Princeton University Press, 1984. Translation by W. D. Ross. Oxford: OCT, 1950.

The Works of Aristotle. Vol. I, *Categoriae and De Interpretatione, Analytica Priora,* and *Analytica Posteriora,* by Aristotle. Translated by E. M. Edghill, A. J. Jenkinson, and G. R. G. Mure. London: Oxford University Press, 1928.

Augustine, Saint, Bishop of Hippo. Sancti Aureli Augustini *Retractationum libri duo,* PL 32, 583-656.

Berkeley, George. *Principles of Human Knowledge.* Edited by Jonathan Dancy. Oxford, Oxford University Press, 1998.

Galilei, Galileo. *Il Saggitore* (The Assayer; 1623). In *Introduction to Contemporary Civilization in the West* Vol. I, 2nd ed. Translation by A. C. Danto. New York: Columbia University Press, 1954.

Hume, David. *An Enquiry Concerning Human Understanding* IV, Part I in *Hume: Theory of Knowledge,* ed. D. C. Yalden-Thomson. New York: Thomas Nelson and Sons, 1951.

A Treaties of Human Nature. Selby-Biggs edition. Oxford, Oxford University Press, 1978.

Kant, Immanuel. *Critique of Pure Reason.* Translation by N. Kemp Smith. London: Macmillan, 1929.

Logic. Translated by Robert S. Hartman and Wolfgang Schwarz. New York: Dover Publication, 1974.

Leibniz, G. W. *New Essays On Human Understanding.* Edited by Peter Remnant, Jonathan Bennett, Karl Ameriks, and Desmond M. Clarke. Cambridge: Cambridge University Press, 1996.

Locke, John. *An Essay Concerning Human Understanding.* Edited by A. C. Fraser (Oxford, 1894).

Suarez, Francisco, S. J. *On Creation, Conservation, & Concurrence: Metaphysical Disputations 20-22.* Translation and introduction by A. J. Freddoso. South Bend, Indiana: St. Augustine's Press: 2002.

Secondary Texts

Adams, Marilyn McCord. *William Ockham* Vol.I. Notre Dame: University of Notre Dame Press, 1987.

Aertsen, Jan A. *Medieval Philosophy and The Transcendentals: the case of Thomas Aquinas.* New York: E. J. Brill, 1966.

Allison, Henry. *Kant's Theory of Taste.* Cambridge: Cambridge University Press, 2002.

Kant's Transcendental Idealism, An Interpretation and Defense. London: Yale University Press, 1983.

Ayers Michael. *Locke.* London: Routledge, 1991.

Balme, D. M. "Greek Science and Mechanism I: Aristotle on Nature and Chance." *Classical Quarterly,* Vol. 33 (1939), 129-38.

Blakemore, S., P. Fonlupt, M. Pachot, C. Darmon, P. Boyer, A. Meltzoff, C. Segebarth, and J. Decety. "How the brain perceives causality: an event-related fMRI study." *Neuroreport* 12 (2001): 3741-3746.

Boehner, Philotheus. *William of Ockham: Philosophical Writings,* Hackett, Indianapolis: 1990.

Bonnette, Dennis. *Aquinas' Proofs for God's Existence, St. Thomas Aquinas On: "The Per Accidens Necessarily Implies the Per Se."* Netherlands: Martinus Nijhoff, 1972.

Brentano, Franz. "Nous Poiētikos." *In Essays on Aristotle's De Anima.* Oxford: Clarendon Press, 1992.

Buersmeyer, Keith. "Aquinas on the 'Modi Significandi'." *The Modern Schoolman* (March 1987): 73-84.

Castelli, E., F. Happe, U. Frith., and C. Frith. "Movement and mind: a functional imaging study of perception and interpretation of complex intentional movement patterns." *Neuroimage* 12 (2000) 314-325.

Casullo, Albert. *A Priori Justification.* Oxford: Oxford University Press, 2003.

Christoff, K., V. Prabhakaran, J. Dorfman, Z. Zhao, J. Kroger, K. Holyoak, and J. Gabrieli. "Rostrolateral prefrontal cortex involvement in relational integration during reasoning." *Neuroimage* 14 (2001): 1136-1149.

Clarke, W. Norris S.J. "Reflections on John Deely's Four Ages of Understanding." *International Philosophical Quarterly* Vol 43, No. 4, Issue No. 172, (December 2003): 527-537.

The One and the Many, A Contempoary Thomistic Metaphysics. Notre Dame: University of Notre Dame Press, 2001.

Courtenay, William J. "The Academic and Intellectual World of Ockham." In *The Cambridge Companion to Ockham*, ed. Paul Vicent Spade, 28-29. Cambridge: Cambridge University Press, 1999.

Deely, John. "Locke's Proposal For Semiotic and the Scholastic Doctrine of Species." *The Modern Schoolman* LXX (1993): 165-187.

Deferrari, Roy J. *A Lexicon of Saint Thomas Aquinas.* U.S.A: Catholic University of America Press, 1948 and 1949.

Evans, Jonathan. "Boethius on Modality and Future Contingents." *American Catholic Philosophical Quarterly*, vol. 78, no. 2 (Spring 2004).

Fabrio, Cornelio. "Knowledge and Perception." *The New Scholasticism* XII (1938): 337-365.

Fonlupt, Pierre. "Perception and judgment of physical causality involve different brain structures." *Cognitive Brain Research* vol. 17/2 (Nov. 2003): 248-254.

Franz, Edward Quinlisk. *The Thomistic Doctrine on the Possible Intellect.* Washington D.C.: CUA Press, 1950.

Freddoso, Alfred J. "The Necessity of Nature." *Midwest Studies in Philosophy* XI (1986): 223.

Frede, Dorothea. "Aquinas on Phantasia" in *Ancient and Medieval Theories of Intentionality.* Dominik Perler (eds.). Leiden: Brill, 2001.

"The Cognitive Role of Phantasia in Aristotle" in *Essays on Aristotle's De Anima.* Martha C. Nussbaum and Amelie Oksenberg Rorty (eds.). Oxford: Clarendon Press, 1997.

Gardeil, H. D., O.P. *Introduction to the Philosophy of St. Thomas Aquinas III*: *Psychology.* Translated by John A. Otto, Ph.D. London: B. Herder Book Co., 1956.

"First Principles" and "Transcendentals." From *Introduction to the Philosophy of St. Thomas Aquinas, Volume IV. Metaphysics.* St. Louis: B. Herder Book, Co., 1967.

Garrigou-Lagrange, R., O.P. *God His Existence and His Nature A Thomistic Solution of Certain Agnostic Antinomies*, London: B. Herder Book Co., 1934.

Gilson, Etienne. *Being and Some Philosophers.* Canada: PIMS, Toronto, 1952.

The Christian Philosophy of St. Thomas Aquinas. Translated by L. K. Shook. New York: Random House, 1956.

Elements of Christian Philosophy. New York, 1960.

Ginsborg, Hannah. "Lawfulness without a Law." *Philosophical Topics* 25, 1997.

"Thinking the Particular as Contained under the Universal" Draft. Dept. of Philosophy, U. C. Berkeley, (November 2003): 1-31.

Goel V., B. Gold, S. Kapur, and S. Houle. "The seat of reason? An imaging study of deductive and inductive reasoning." *Neuroreport* 8 (1997): 1305-1310.

Goodenough, Judith, Robert A. Wallace, and Betty McGuire, *Human Biology*. Fort Worth: Saunders College Publishing; Harcourt Brace College Publishers, 1998.

Haldane, John J. "Aquinas on Sense-Perception." *The Philosophical Review,* XCII no. 2 (Ap. 1983): 233-239.

Henle, Robert J. "Apropos of McCool." In *Thomistic Papers VI*, 144-145. Texas: The Center for Thomistic Studies, 1994.

Theory of Knowledge. Chicago: Loyola University Press, 1983.

Henrik Lagerhund. "Section 4 Peter Abelard" of "Medieval Theories of the Syllogism." In the *Stanford Encyclopedia*; available from http://plato.stanford.edu/entries/medieval-syllogism/; Internet; accessed 19 February 2005.

Hintikka, Jaakko. "Aristotle on Modality and Determinism." *Acta Philosophica Fennica* Vol. 29, No. 1 (1977): 13-22.

Knowledge and Belief: An Introduction to the Logic of the Two Notions. NY: Cornell University Press, 1962.

Models for Modalities. Dordrecht: Reidel, 1969.

"Necessity, Universality and Time in Aristotle." *Ajatus,* 20 (1957): 65-90.

Time & Necessity. Oxford: The Clarendon Press, 1973.

Time and Necessity: Studies in Aristotle's Theory of Modality. Oxford: Oxford University Press, 1972.

"The Varieties of Being in Aristotle." In *The Logic of Being: Historical Studies*. Editors S. Knuuttila and J. Hintikka. Dordrecht: Reidel, 1986.

Hintikka, J,. Remes, U. and Knuuttila, S. *Aristotle on Modality and Determinism*. Amsterdam: North Holland, 1977.

Humber, James and Edward Madden. "Natural Necessity." *The New Scholasticism* vol. XLVII, no. 2 (Spring 1973): 214-227.

Jones, W.T. *Kant and the Nineteenth Century.* New York: Harcourt Brace & Company, 1980.

Jaworski, William. "Hylomorphism and the Mind-Body Problem." Paper delivered at the American Catholic Philosophical Association Annual Meeting November 5-7, 2004.

Klima, Gyula. "Medieval Problems of Universals." In the *Stanford Encyclopedia of Philosophy,* available from www.seop. leeds.ac.uk/entries/universals-medieval/; Internet; accessed 19 February 2005.

"The Semantic Principles Underlying Saint Thomas Aquinas's Metaphysics of Being." *Medieval Philosophy of Theology* 5 (1996): 87-141.

Klubertanz, George P., S.J. *The Discursive Power.* Ohio: The Messenger Press, 1952.

"The Internal Senses in the Process of Cognition." *The Modern Schoolman* (January 1941): 27-31.

"St. Thomas and the Knowledge of the Singular." *The New Scholasticism* XXVI (1952): 135-163.

Knuuttila, Simo. "Medieval Theories of Modality." In the Stanford Encyclopedia; available from http://plato.stanford.edu/entries/modality-medieval; Internet; accessed 19 February 2005.

Modalities in Medieval Philosophy. New York: Routledge, 1993.

"Time and Modality in Scholasticism." In S. Knuuttila ed., *Reforging the Great Chain of Being: Studies of the History of Modal Theories.* Dordrecht: Reidel, 1981.

Kossel, Clifford G., S. J. "St. Thomas's Theory of the Causes of Relation." *The Modern Schoolman* vol. 25, no. 3 (March 1948): 155-167.

Kripke, Saul. "Identity and Necessity." *In Metaphysics Contemporary Readings*. Editor Michael J. Loux. London and New York: Routledge, 2001.

Naming and Necessity. Massachusetts: Harvard University Press, 1980.

Kuehn, Manfred. "Kant's Critique of Hume's theory of faith." *Hume and Hume's Connexions*, ed. Stewart and Wright. Pennsylvania: Pennsylvania State University, 1995.

Lagerliund, H. *Modal Syllogistics in the Middle Ages*. Leiden: Brill, 2000.

"Medieval Theories of the Syllogism." In the Stanford Encyclopedia; available from http://plato.stanford.edu/entries/medieval-syllogism; Internet; accessed 19 February 2005.

Laumakis, Stephen John. "The 'sensus communis' and the Unity of Perception according to Saint Thomas Aquinas." Ph.D. Dissertation. Notre Dame: University of Notre Dame, 1991.

Lisska, Anthony J. "Thomas Aquinas on Phantasia: Rooted in But Transcending Aristotle's De Anima." Available from www.nd.edu/Departments/Maritain/ti00/lisska.htm; Internet; accessed 19 February 2005.

Lokhorst, Gert-Jan C. "Philosophy and the Brain." Paper delivered at the University of Helsinki, May 14-15 2001 (transparencies available from http://www2.eur.nl/fw/staff/lokhorst/helsinki.2001.html; Internet; accessed 19 February 2005).

Lombardi, Petri. Dionysius' *Liber De causis*. In *Scriptum super libros magistri Petri Lombardi*, vol. 1, d. 38, q. 1, a. 2, ed. R. P. Mandonnet. Paris: P. Lethielleux, 1929.

Lonergan, Bernard. "St. Thomas's Theory of Operation." *Theological Studies*, III (September, 1942): 375-383.

Longuenesse, Beatrice. *Kant and the Capacity to Judge*. Translated by Charles T. Wolfe. Princeton: Princeton University Press, 1998.

MacIntosh, J. J. "Aquinas on Necessity." *American Catholic Philosophical Quarterly*, vol. LXXII (1998): 374-378.

Mahoney, Edward P. "Aquinas's Critique of Averroës' Doctrine of the Unity of the Intellect." In *Thomas Aquinas and His Legacy*, ed. David M. Gallagher. Washington D.C: The Catholic University of America Press, 1994.

"Sense, Intellect and Imagination in Albert, Thomas, and Siber." *In The Cambridge History of Later Medieval Philosophy*. Edited by Norman Kretzmann, Anthony Kenny and Jan Pinborg, (Cambridge: Cambridge University Press, 1982, 1984): 602-622.

Maritain, Jacques. *Formal Logic*. New York: Sheed & Ward, 1946.

An Introduction to Philosophy. New York: Sheed & Ward, 1937.

he Degrees of Knowledge. Indiana: University of Notre Dame Press, 1995.

Martin, R. M. *Logic, Language, and Metaphysics*. New York: New York University Press, 1971.

Maurer, Armand A. *Being and Knowing: Studies in Thomas Aquinas and Later Medieval Philosophers*. Toronto Ontario: Pontifical Institute of Mediaeval Studies, 1990.

McInerny, Ralph. *Being and Predication*. Washington, D.C.:Catholic University of America Press, 1986.

Meehan, F. X. *Efficient Causality in Aristotle and St. Thomas* Ph.D. dissertation. Washington, D. C.: Catholic Univ. of America Press, 1940.

Meyer, Hans. *The Philosophy of St. Thomas Aquinas.* Translation by Rev. Frederic Eckhoff. St. Louis: B. Herder Book Co., 1944.

Michalson, Gordon E. Jr. *Lessing's 'Ugly Ditch': A Study of Theology and History.* United States: The Pennsylvania State University, 1985.

Miller, Kenneth R. and Joseph Levine. *Biology.* New Jersey: Prentice Hall, 1998.

Miller, William E. "Aristotle on Necessity, Chance, Spontaneity," *The New Scholasticism*, vol. XLVII, no. 2 (Spring 1973): 204-213.

Murray, Andrew. *Intentional Species and the Identity Between Knower and Known According to Thomas Aquinas.* Washington, D.C.: The Catholic University of America, 1991.

Owens, Joseph. *St. Thomas Aquinas on the Existence of God.* The Collected Papers of Joseph Owens, ed. by John R. Catan. Albany, N.Y.: State University of New York Press, 1980.

Pierre Fonlupt. "Perception and judgment of physical causality involve different brain structures," *Cognitive Brain Research* vol. 17/2 (Nov. 2003): 248-254.

Pippin, Robert. *Kant's Theory of Form.* New Haven: Yale University Press, 1982.

Prior, A. N. "Can religion be discussed?" *New essays in Philosophical theology*, ed. A. G. N. Flew and A. C. MacIntyre (1955): 1-11

Radcliffe, Elisabeth. *Proceedings and Addresses of The American Philosophical Association.* DE: The American Philosophical Association, 2002.

Rocca, Greggory, O.P. "Res Significata and Modus Significandi", *Thomist* , 55(1991): 177180.

Ryan, Edmund J. *The Role of the "sensus communis" in the Psychology of St. Thomas Aquinas.* Cartheena, Ohio: The Messenger Press, 1951.

Ryan, John K. "Aquinas and Hume on the Laws of Association," *The New Scholasticism*, vol. XII, no. 4 (October, 1988): 366-371.

Schlottmann, A., and D.R. Shanks. "Evidence for a distinction between judged and perceived causality." *Q. J. Exp. Psychol.* 44 (1992): 321-342.

Schmidt, Robert W., S.J. *The Domain of Logic According to Saint Thomas Aquinas*. Netherlands: Martinus Nijhoff, 1966.

Sellars, Wilfred. "Empiricism and the Philosophy of Mind." In Science, Perception and Reality. London: Routldge and Kegan Paul, 1963.

Shook, Lawrence K. *Etienne Gilson*. Toronto: Pontifical Institute of Mediaeval Studies, 1984.

Sorabji, Richard. "Aristotle on Sensory Processes and Intentionality. A Reply to Miles Burnyeat" in *Ancient and Medieval Theories of Intentionality*, Dominik Perler (eds.). Brill, Leiden: 2001.

Starr, Cecie. *Biology concepts and applications*. United States: Wadsworth Publishing Company, 1997.

Sullivan, Mark W. *Apuleian Logic The Nature, Sources, and Influence of Apuleius's Peri Hermeneias*. Amsterdam: North-Holland Publishing Company, 1967.

Suarez, Francisco S.J. *On Creation, Conservation, and Concurrence*. Translation, notes, and introduction by Alfred J. Freddoso. South Bend, Indiana: St. Augustine's Press, 2002.

Suto, Taki. "Boethius on Mind and Language: For a Study of Boethius' Commentary on Peri hermeneias" (available from http://www.hmn.bun.kyoto-u.ac.jp/report/2_tetsugakui/2_16.pdf; accessed 19 February 2005).

Swedenborg, Emanuel. *Psychologica, Being Notes and Observations on Christian Wolff's Psychologia Empirica*. Philadelphia: Swedenborg Scientific Association, 1923.

Sweeney, Leo. "Must Thomism Become Kantian." In *Thomistic Papers VI*, 172-194. Texas: The Center for Thomistic Studies, 1994.

Thomas, Dom and Verner Moore. "Gestalt Psychology and Scholastic Philosophy." *The New Scholasticism* VII (1933): 298-325m; VIII (1934): 46-80.

Wallace, William A., O.P. *The Elements of Philosophy*. New York: ALBA House, 1977.

Watkins, Eric. *Kant and the Metaphysics of Causality*. Cambridge: Cambridge University Press, 2005.

Weiten, Wayne. *Psychology* 4th ed. Mexico: International Thomson Publishing Inc., 2000.

Westphal, Kennith R. *Kant's Transcendental Proof of Realism*. Cambridge: Cambridge University Press, 2004.

Wippel, John F. "Thomas Aquinas on Our Knowledge of God and the Axiom that Every Agent Produces Something Like Itself", *Philosophical Theology: Reason and Theological Doctrine*, vol. 74 in the *Proceedings of the American Catholic Philosophical Association*, ed. Michael Baur (NY: Fordham University, 2000): 84-90.

White, Alfred Leo. *The Experience of Individual Objects In Aquinas A Dissertation*. Washington, D.C.: The Catholic University of America, 1997.

Wilkerson, T. E. *Kant's Critique of Pure Reason, A Commentary for Students*. Oxford: Clarendon Press, 1976.

Wippel, John F. *Metaphysical Themes in Thomas Aquinas*. Washington, D.C.: Catholic University of America Press, 1978.

Wolff, Robert Paul. "Kant's Debt to Hume via Beattie." *Journal of the History of Ideas* 21 (January-March 1960): 117-123

Kant's Theory of Mental Activity. Gloucester, Mass.: Peter Smith, 1973.